POLLUTION CONTROL COSTS IN INDUSTRY:

AN ECONOMIC STUDY

Other Titles of Interest

CUMES, J. W. C.
Inflation! A Study in Stability

GOLD, Bela
Technological Change: Economics, Management and Environment

IBRAHIM, I. B. et al.
Readings in Managerial Economics

LETHBRIDGE, D. G.
Government and Industry Relationships

MELNYK, M.
Principles of Applied Statistics

SANDFORD, C. T.
Economics of Public Finance, 2nd edition

STILWELL, F. J. B.
Normative Economics: An Introduction to Microeconomic Theory and Radical Techniques

POLLUTION CONTROL COSTS IN INDUSTRY:

AN ECONOMIC STUDY

by

M. H. ATKINS
Trent Polytechnic, Nottingham

and

J. F. LOWE
University of Bath

PERGAMON PRESS
OXFORD · NEW YORK · TORONTO · SYDNEY · PARIS · FRANKFURT

U.K.	Pergamon Press Ltd., Headington Hill Hall, Oxford OX3 0BW, England
U.S.A.	Pergamon Press Inc., Maxwell House, Fairview Park, Elmsford, New York 10523, U.S.A.
CANADA	Pergamon of Canada Ltd., 75 The East Mall, Toronto, Ontario, Canada
AUSTRALIA	Pergamon Press (Aust.) Pty. Ltd., 19a Boundary Street, Rushcutters Bay, N.S.W. 2011, Australia
FRANCE	Pergamon Press SARL, 24 Rue des Ecoles, 75240 Paris, Cedex 05, France
WEST GERMANY	Pergamon Press GmbH, 6242 Kronberg-Taunus, Pferdstrasse 1, West Germany

---

First edition 1977

Library of Congress Catalog Card No. 77-78736

Printed in Great Britain by Butler and Tanner Co. Ltd. Frome.

ISBN 0 08 021851 2 (Hard cover)
ISBN 0 08 021841 5 (Flexicover)

To Dennis, John and Ron —
an unlikely trio . . .

CONTENTS

PREFACE

A major problem of contemporary economic and social policy is the extent to which industrial pollution should be controlled. To this end a substantial amount of academic research has been carried out across the whole spectrum of science. Much theorising and research has still to stand the test of time since environmental issues are essentially long term and there is considerable uncertainty in many natural and applied sciences about pollution and its control.

In the sphere of economics, substantial advances have been made at the macro level concerning problems such as economic growth and the quality of life, but there are still many under-researched areas at the micro level. Little work has been done outside the US on the impact of pollution controls on firms and industries in general, and small and medium sized firms in particular. This book sets out to fill part of this gap. The first chapters deal with some important strands of economic theory underlying the problem of environmental pollution and later chapters analyse the results of over one hundred interviews we had with firms on their pollution problems and means of abatement.

Talking to firms about their resource allocation and day-to-day decisions is an unfashionable aspect of industrial economics, but one which we feel is nevertheless important, particularly in the area of this study. Without this, 'micro' information policies to control pollution cannot be framed adequately.

This book is intended primarily for undergraduate Economics and Business Studies students interested not only in environmental issues but also in industrial organisation and the theory of the firm. It is also hoped that students in pure and applied science might find the text a useful introduction to some of the economic issues of environmental policy.

The bulk of the research was carried out whilst the authors were lecturers in the Department of Industrial Economics at the University of Nottingham, and we are indebted to our colleagues there for their helpful comments and suggestions. Particular thanks are due to Keith Bowker who carried out many of the interviews with local authorities and firms. Also to Ron Arnfield who was always a great source of help and ideas, particularly in the initial stages of the project. To Dennis Lees, whose belief in the importance of empiricism in economics was a constant source of guidance and inspiration. To John Butterfield who always showed an interest and encouraged us in our task. Finally, thanks are due to the financial assistance of Allied Breweries, for without their generosity this work would not have been possible.

Any errors or omissions are, of course, the sole responsibility of the authors.

INTRODUCTION

The Nature of Pollution

Pollution can be defined as the production of 'bads' rather than goods. Pollution may have a visual impact, a smell, a sound or noise, or may be defined in terms of its physical characteristics; it may be something which is solid, liquid or airborne and which causes some discomfort or harm to at least some sections of the population. Many so-called pollutants are readily apparent and their effects well documented yet may not always cause harm in a directly physical sense and may even have an aesthetic appeal for some; others may exist unknown either to the general public at large, as a result of a poor communications system, or unknown to those who are directly concerned with their manufacture because of insufficient knowledge of the processes involved. This very broad definition of pollution is likely to lead to a large number of phenomena being classified as pollutants.

Generally we think of pollutants arising out of production but clearly they can also arise out of consumption - from fumes from driving a car to the sound pollution of someone rustling a bag of crisps in the theatre. Pollutants have clearly been with us since time immemorial and have, to a large extent been tolerated as a fact of life since, if the production of all 'bads' were prevented there would be precious little production or consumption of goods. To a certain extent the increases in living standards in Western economies which have occurred during the 19th and 20th centuries have merely exacerbated the problem and pollutants that have always existed have become a much more explicit feature of life. Not only has the scale of production of 'bads' increased but their nature has changed for the worse; for example, toxic wastes that the environment is incapable of destroying such as DDT. Coupled with this, education and the mass media have generated information on pollution such that individuals are much more concerned with the issues that are collectively referred to as environmentalism. Yet this increased public awareness of the 'bads' that affect our daily lives and the desire to eliminate them does not always explicitly admit that as producers and consumers we all produce some 'bads' and that their prevention would not always be desirable or practicable.

Economists have traditionally considered the external effects of production and consumption and in a seminal article R. Coase [1] explicitly examined the reciprocal nature of pollution. Specifically, it was suggested that the effects of pollution prevention could potentially be just as harmful to society as allowing pollution to continue. Environmental protection has to be paid for by society so, to an extent, some pollution may be a good thing. The scope of control is, however, a problem economists cannot deal with in isolation since the definition 'goods' and 'bads' is dependent upon the individual's viewpoint and the desirability of a distribution of wealth which would allow the purchase of goods and an environment free from bads is something which would have to be decided through the ballot box. Whatever the result, it seems inevitable that conflict must exist and this conflict found explicit recognition in the late 1960's amongst those (see for example 2 and 3) who felt that the pursuit of growth, in the form of increases in national income, is a necessary and desirable aspect of society. Indeed, viewing society's welfare in the aggregate this is probably so, since the demand for goods arises because they represent potential increases in welfare and the consumption of bads is not often perceived as representing a decrease in welfare. The Galbraithian view of the corporate state [4] dominated by large firms which create demand rather than react to it, would of course suggest that goods and bads are joint products of some fait accompli. If, however, we look at society's welfare from the individual's viewpoint, whether or not economic growth leads to an improvement in welfare depends on the distribution and composition of any increase in national income. Other commentators (such as 5) have looked at the question as part of the failure of the market economy; the problem is that firms use certain 'free' resources like air and tranquility or resources that are under-priced like water and roads. The fact that a price is arrived at which is less than a perfectly competitive market price ensures that there will be excess demand and over-usage of these resources. Previous empirical economic studies (some of which have been discussed in 6) have been devoted to topics such as the pricing and use of water resources or the impact of a new road on the economic welfare of individuals (as judged in terms of house prices, leisure and work opportunities and accident costs).

Generally then, economic studies of pollution have related to the impact of the polluter on the polluted, even though there is compelling logic for an approach in the reverse direction.

The Research

A priori there appear good reasons for looking at the impact on firms of government controls and public awareness of pollution. Specifically, these may result in costs the firm has to bear, directly as a result of the emission of a waste product, or alternatively the firm may be prevented from using an input that, under other circumstances, it would choose to use. The prevention of firms using certain inputs and the charging for waste outputs necessarily constrains freedom of choice and may affect the efficiency of the firm at that point in time; this may result in an increase in prices, reduction of employment or reduction in product quality. The extent to which these phenomena occur depends on the relative importance of pollution control costs in production. The size and nature of the polluter's output of waste also has to be determined, although sometimes it may be necessary to use inputs as a proxy indicator of outputs, for instance when trying to quantify airborne emissions. This type of data is not available from published sources and consequently it had to be collected on an individual firm basis through the use of a questionnaire and interview to generate the information. We made no assumptions regarding which firms were polluters and instead began with a broad-based sample to eliminate any such subjective bias from our research. We were not even sure of the type of data firms would have or were willing to disclose. Thus a major objective of the research was one of finding out what was needed to be found and what sort of relationships between pollution and individual firms exist.

Manufacturing industry in the East Midlands region represented the basis of our study and, since we did not wish that any one firm or organisation should dominate our research, it was decided to limit it to small and medium sized firms (defined by 0-200 and 200-1000 employees respectively). The choice of geographic area for our sample was, apart from considerations of local interest, a useful one since the region is of average prosperity relative to the rest of the UK and has a well-diversified manufacturing sector (this is elaborated upon in Chapter III). Since we were looking at small and medium sized firms, it appeared likely that we would encounter the whole range of managerial motivations, from the small firm run by the owners, whose prime interest *prima facie* would probably be pecuniary gains, to the medium sized firms run by professional managers on the behalf of shareholders, where the motivation of the former could clearly diverge from the straight profit motive. Economic theory has often concerned itself with the impact on resource allocation within the firm of differences in ownership and control; the study of pollution control represented another specific issue of a general area of research into these relationships. Since our intention was to generate information on an individual firm basis through questionnaire and interview, another important consideration in selecting the firms for study was ease of information collection and interpretation. The fact that in firms like these the division of responsibility is limited relative to the very large corporation, and the connections between various profit centres and the overall economic health of the firm are relatively simple, should minimise these potential data problems.

Some Conceptual Issues

Since our major objective was to decide how pollution control affected the activities and health of firms, the first conceptual step was to define pollution. In the broader context of the economy, we have already pointed out that pollution is sometimes a very subjective phenomenon and can perhaps best be described as the 'bads' emanating from production and consumption of 'goods'. However, there are many 'bads' that people have learnt to live with or have not perceived as 'bads'. The problem within the firm is just the same and, since we did not wish to become involved in the tricky and technically demanding process of measuring when a pollutant is a 'bad' as defined by some arbitrary standard, we decided that for the purposes of the investigation anything that was considered a 'bad', either by managers, workers, or government representatives, would be relevant for our study. The usefulness of this approach can be seen when we investigated noise, for which definitions and measurements are legion. From the viewpoint of our study it was merely something which, to some, represented a 'bad'. The second conceptual problem was to define when a cost could be attributed to pollution control as opposed to a change in technology *per se*. Sometimes, of course, as we found later, pollution control could lead the firm to more efficient and less polluting means of production. Strictly, pollution control costs would be counted as such when their objective was the dealing and processing of a non-saleable waste product.

Whilst a primary objective of the study was data collection and appraisal to serve as a basis for future work, we had some broad notions about potentially interesting areas of investigation. In particular we expected pollution control would be a relatively new and rapidly increasing cost the firm has to pay. This could potentially be levied unevenly, arbitrarily and sometimes unfairly, between products within a firm, between firms and industries in a region, between regions and even between countries. *A priori* then, it seemed likely that there would be competitive and market structure changes because of the imposition of these controls and possible unevenness in their application. For instance, if pollution control represented a heavy *fixed* cost element that did not vary greatly with output, it might be expected that 'bigness' would be favoured and this might lead to mergers and closures in some industries.

The incidence of pollution costs (that is, whether the cost is passed backwards to share-holders and labour or forwards to consumers) represents another interesting area of research. There were clear a priori reasons for thinking that product price or quality, or firm employment might be affected by the levying of an effluent charge or prevention of a process. The main problem here was information collection and the interpretation of relationships. Our study was bascially cross-sectional but often when technological innovation, for instance, was the ultimate result of pollution enforcement it was clear that a time-series approach would be more useful. Finally we felt that a profitable area of study of something as subjective and nebulous as pollution was the transmission and perception of information within the organisation - between managers and managers, managers and workers, and workers and workers - and between the firm and the authorities. Information transmission and perception could well be a key issue for firms, their representatives and public bodies.

Our ultimate goal, then, was to gain a better understanding of the impact of pollution on society looked at from the viewpoint of industry and, if possible, to estimate the total resource costs of pollution control for industry in the East Midlands.

References

1. R. Coase, The Problem of Social Cost, Journal of Law and Economics, 1960.

2. W. Beckerman, Why We Need Economic Growth, Lloyds Bank Review, October, 1971.

3. E.J. Mishan, The Costs of Economic Growth, Pelican Books, 1967.

4. J.K. Galbraith, The New Industrial State, Houghton Mifflin, 1967.

5. R. Turvey, On divergences between private and social cost, Economica, 1963.

6. R. Millward, Public Expenditure Economics, McGraw-Hill, 1971.

Chapter I

THE ECONOMICS OF POLLUTION

Externalities: The General Case

Economists' interest in the problems of pollution stems principally from three sources: firstly, the existence of externalities; secondly, the debate concerning the costs and benefits of economic growth and finally, the incidence of pollution control on the firm, the consumer and labour. The essence of the debate on externalities (the divergence between private and social costs) has been that production and sometimes consumption of goods uses scarce resources, some that can be bought (labour, raw materials, etc.)and some that result from common property for which, in most instances, no price is paid (for example fresh air and clean water), since they are used by the community as a whole and there is no market in them. Two problems arise because of this state of affairs; (i) there is the possibility of over-usage * and (ii) because the resources are common property, the consumption of one section of the community may well affect the welfare or satisfaction of another group, thus the smoky chimney pollutes the air others breath and the chemical effluent kills the fish in the river or, indeed, one firm's pollution may add to another firm's production costs. Technically this means that the individual's utility function contains arguments, some of which he has under his control and some that are inter-dependent with the production or consumption processes of others. The main problem that economists have attempted to solve in this area relates to what amount of pollution is consistent with an optimum allocation of resources and how can this optimum state of affairs be arrived at. Closely connected to these issues, which have typically been investigated at the micro level (firm, individual or industry), is the debate over whether or not economic growth is a desirable macro-economic objective and exactly how the level of national output can be calculated to take account of non-market activities (like some forms of leisure) which rely on the use of common property resources and do not command a market price. Finally pollution abatement measures are taxes on the firm's operations and affect the level and nature of these operations to some extent; how the firm reacts to these controls and what impact they ultimately have on the firm's prices, output, employment and technical progress represents another important area where economics is able to make some contribution. Our research has been concerned mostly with the micro level so we present below a brief overview of the economic theory relating to the first and last subject areas listed above.

Normally, theory has concentrated on a two-party model consisting of the gainer, who pollutes and causes the externality, and the loser, who suffers from the pollution. Policy proposals may involve one or both parties changing their operations, which will be

---

* This word is used in a technical sense to define a state of affairs that arises when more of an input is consumed than would have been if the user had to pay some realistic positive price.

expected to lead to a change both in the distribution of income and the technical allocation of resources. This is a crucial distinction, since the arguments about the efficiency impact of pollution centre primarily on the effect on Paretian efficiency * and the technical allocation of resources is quite a separate issue from that of justice and the distribution of income. This distinction between distributive and allocative efficiency has led to some incredible pronouncements on how to deal with pollution offenders in order to ensure that their activities are congruent with a Paretian efficient allocation of resources. To illustrate, we can assume perfect competition and proceed with a partial equilibrium analysis of two parties, one who is the polluter (whom we call G for gainer, the owner of a smoky chimney) and the other who suffers the pollution (L for loser, a local resident). Let us assume that the nature and size of the externality will depend on the scale of G's activity, thus the more iron castings our foundry produces, the greater is the gain to the founder and the greater is the loss to the smoke and grit polluted resident located nearby. This loss to the resident we assume can be financially quantified in some way, as can the gain to the polluter. We assume that the gain to the producer increases at a diminishing rate and that, as his level of activity increases, the loss to the polluted will increase at an increasing rate, thus we can represent the marginal gain and marginal loss to the polluter and polluted as in Fig. 1.1.

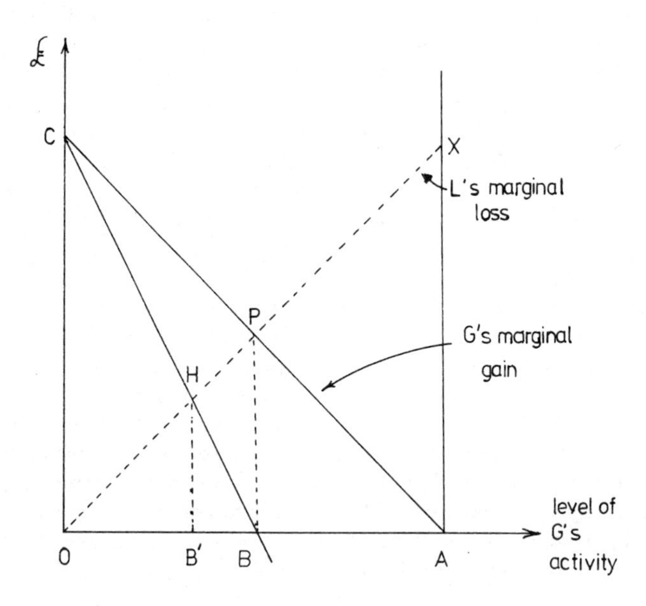

Fig. 1.1

* A Pareto optimal solution is said to exist when further changes in production and exchange conditions cannot further increase the total satisfactions of society.

Now, if G is a profit maximiser we assume that in the absence of constraints he will extend his polluting activity to OA where his marginal gain is zero. At this level of activity, L suffers a marginal loss of XA and a total loss of OXA. Since his loss per unit of output is greater than G's gain, then we would expect that to concur with the Pareto optimal solution marginal gains and losses of both G and L should be the same, thus the optimal scale of G's activity is at B, but it must be remembered that this is optimal only from an allocative and not a distributive point of view. The solution public decision-makers were seeking could be achieved by levying a tax on G which varied exactly with the losses he imposed on L. Thus the polluter's marginal gain curve including taxation to the level of his losses is given by the line CB which means in isolation G's marginal gain would equal zero at B, which as we have just shown is the Pareto optimal level of his activity.

Two main arguments have been advanced against this Pigovian doctrine of taxing the polluter to the extent of the damage he causes. The first is that the achievement of OB of G's activity would result in any case, because past this point G's gains are less than his losses and consequently L would want to bribe G to reduce the scale of his activity. He could continue to do this as long as L's losses exceeded G's gains and this would be up to and including point B. Thus a tax to achieve this solution would be irrelevant; the argument continues, however, to suggest that the tax would be positively harmful since it would in fact create a point to the left of B (the Pareto optimum point) which would equate L's losses and G's net of tax gain. Thus a rational L would be able to bribe G to reduce activity back to point B^1 since, at that point, the marginal gains and losses are equal. Some economists would then view a policy of fiscal intervention to control pollution as something which distorts the allocation of resources in a free market. There is, however, a suitable repost since the arguments of this particular 'lunatic fringe' (1) would, it is presumed, suggest that the victims of criminal attacks should bribe prospective attackers to leave them alone and, if this appears to be an argument which still depends on value judgments as to who has the right in a free society to do what, then it can still be shown that the tax would not in fact lead to a sub-Pareto point B^1 since the recipients of the tax revenue (let us call them C) would prefer that G's activity was at B rather than B^1 as this represents higher tax accruals. This other group would then be prepared to bribe G to move back to some point nearer to B than B^1. In the real world this doctrine of losers bribing gainers might seem quite ludicrous since frequently the losers are a heterogeneous group of individuals and the costs of getting them to act together might exceed the benefits of so doing. A tax might then be necessary to compensate for a system which has positive contact, contract and administration costs. However, the bribery approach does highlight one very important factor which may previously have been ignored; when we are dealing with free resources which everyone may use, if one user has to reduce consumption because of the impact he has on the other user, he is being subjected to costs because of another's preference. Externalities are thus basically reciprocal :- the nearby residents who breathe the smoke from a dirty chimney must share responsibility for the resulting social cost. True, the social cost would disappear if the factory ceased production but the same is true if the residents moved away. Our investigations of the way in which local authorities controlled pollution suggested that the people most likely to complain of noxious fumes and other matters were the new residents in an area who moved some time after the first emission took place; this evidence would certainly support those who would suggest that pollution is essentially a reciprocal problem. Thus just as the smoke emitted by the factory imposes at least psychic costs on its neighbours, the latters' insistence on the installation of purification devices or a reduction of pollution activity imposes a cost on the factory and, depending on the incidence, the rest of society. If it is socially less costly to move the polluted than stop the pollution, a tax which adopts the latter course may result in resource misallocation. However, economic policy will ultimately depend on value judgments being made about these issues where there are no clear cut property rights. Where value judgments are made by society a Pareto optimum point need not necessarily represent a position of welfare maximisation since welfare maximisation will include distributional value judgments as well as decisions concerning technical efficiency. The theoretical debate over whether government should intervene, or whether in fact the market place has its own allocative efficient solutions, continues with an enormous output of paper but little of real policy relevance. However, the arguments for government intervention are convincing and the growing concern and powers of the executive have been reflected in the UK in various items of legislation (2) which are aimed at providing guidelines for local and central government to control the polluting activities of industry and individuals. They generally admit to the principle of "the polluter must pay" or "the polluter must be stopped" and it is the choice between direct physical controls through the taxing of polluters to which we now turn.

The theoretical debate is quite straightforward and the most influential economists all seem to point in the direction of charges being preferable to regulation, yet in the UK and elsewhere regulated standards are common, especially in combating air and noise pollution, and are at least as important as charges in the control of liquid pollution. It seems that what is preferred in theory is least preferred in practice; to understand why this is so we must first examine the theory*. The theoretical arguments are normally conducted in a world of zero information, administrative and legal costs and an objective of welfare maximisation rather than merely some mutually satisfactory state of affairs. In the idealised world with perfect information the policy maker is in a position to know the cost and revenue function of the firm as well as the damage function incurred by the environment. The previous analysis of externalities demonstrated that a pollution charge equal to the value of the external costs leads to an allocative efficient solution by each firm; however, it is clear that armed with this same information the policy maker could induce the firm to achieve this optimum at the same level of output by means of regulatory control. If the optimum output levels differ between firms, a set of standards will be needed. Generally in the idealised model the two forms of regulation are equivalent, although it might be argued that the prevention standard does not allow the firm the choice to produce beyond this point, whereas a tax merely makes it unprofitable for it to do so. Firms with non-profit maximising objectives, or firms who had a substantial slack component in their costs, might have a higher level of output and employment when charges prevail rather than when prevention standards are used. Burrows discusses the possibility that prevention standards might lead to high levels of output and employment when there are variable production processes (3). This is due to the fact that some forms of regulated standards allow pollution up to that standard free of charge, whereas pricing internalises the externality at all levels.

Under regulations rather than charges, the result is that the lower marginal cost of production leads to a higher output and may represent a greater uncompensated loss to the environment by society. We therefore need criteria broader than that of welfare economics to determine what is a reasonable level of pollution.

---

* Much of the following discussion is based upon Paul Burrows (3).

In the theoretical world, the welfare optimum is chosen by society from the many different Pareto optima. In the real world, information is costly and Pareto optimal points somewhat difficult to define so, if a discussion of taxation versus direct controls is to have meaning, it must rest ultimately on an objective which is based on some predetermined standard of an acceptable level of pollution which is in essence easily definable and policeable. The major problem with taxation in this situation is that the right tax can only be determined by some iterative procedure and if the tax begins much too low or too high we may get a situation whereby firms invest in new equipment and processes which will be able to work within the fiscal charging system but then have to change at substantial cost later. Taxation under these circumstances could have a very serious impact on the level of production and prevention costs although the evidence of the remainder of this report indicates this to be unlikely. The same problem does not arise if universal standards are applied to all firms but where firms have different prevention costs, it will be necessary to set a separate standard for each firm. This has for a long time been recognised in British legislation under the "best practicable means" clause and by the operations of the complaints procedures adopted by local authorities which ensure that the firms which have to conform most closely to the standards are those close to urban residential developments.

Probably the more convincing arguments in favour of taxation are those associated with the ideas of freedom and incentive. It may be claimed that taxation allows the firm freedom to find the cheapest means of pollution regulation, whether it be through changes in the types of input or changes in the production processes, although as Burrows points out, it is difficult to see why firms under regulation are not just as free to find the cheapest method of pollution reduction. What can happen, of course, especially when there are no effective ways of preventing pollution, is that the consumer ultimately will have the choice of paying more for a product which involves pollution in its manufacture, whereas the regulated industry may just have to stop production. The argument here is analogous to the choice between quotas and tariffs in international trade and since the latter give more choice they must be better on those grounds alone. This is also true of taxation rather than standards although clearly in this case the elasticity of demand will be the crucial factor, whereas we assume that quotas would not be needed unless demand was proving to be relatively inelastic. The other major advantage of taxation is that at up to all levels of production there is an incentive to reduce pollution and this increases with the severity of pollution. In the case of regulations, however, the need to search for new processes will be enhanced when the standard is approached. *A priori* we cannot really come to a conclusion; however, the dynamic incentive to innovate under the taxation system may ultimately be crucial. Finally, one problem with taxation is that, under conditions of rapid inflation and money illusion, consumers and producers may lose track of changes in relative prices which will be disguised if the rates of inflation vary. It may be easy to pass on price increases and society may suffer more pollution than it really wanted to pay for if it knew the real prices involved.

The arguments concerning which system would cost more to operate and administer are just as ambivalent. On the one hand there are heavy costs of finding out each firm's cost and revenue function in order to know what is the tax necessary to reduce his gain to zero at a required level, whilst the iterative procedure referred to before may generate poor or unwarranted investments. Additional to this, there is a transfer of funds from the manufacturing sector to government and this in itself requires real resources to operate.

Possibly with regulation there is a greater necessity for policing, although something like that would be needed under a taxation system to prevent non-payment of fines, etc. On the other hand, there are problems with regulation if more than one standard is required (i.e. if the firms are heterogeneous. It seems clear that no definitive judgment in favour of the cost of either system can be made.

Equity is important too; the availability of low-pollution processes, or low-cost pollution prevention, may in a real market economy be immaterial to allocative efficiency and a single standard should apply. However, the single standard which bears harder on some than others may not be what society considers fair. Even on allocative grounds there may be cause for disquiet because the same quantity of pollutants in two different areas may cause vastly different levels of harm (cf. the urban and rural factory) and thus should not be subject to the same standards. However, if the charge of those supporting tax solution is that a variety of standards are needed and that this would cause administrative chaos, then it must also be true that a variety of taxes would be needed too.

Finally, the pricing solution does depend on the "ability to pay" principle and Burrows feels this may be out of line with our rejection of that principle in areas such as health and education. This is true and may lead from the "polluter must pay" to the "payer may pollute" principle. Freedom from pollution regulations can be achieved by the rich more than the poor, although they do not allow the rich to pollute more than the poor.

Our conclusion as to which system of control is best must ultimately depend on which industry and which pollutant because factors such as information on costs and benefits to the polluted and polluter and policing of the law cannot be generalised for all products and processes, which may indeed be why the two types of regulation exist side by side in the UK today.

In the next section we develop a model that takes account of information constraints while also analysing the impact of production processes, the technical nature of pollutants, and community choice on the aggregate pollution load of a growing economy.

Satiation, Information and the Problem of Wastes*

Most production processes create waste by-products which are disposed of in some way; some command a positive market price and are recycled or used in some other way and others constitute an economically or technically unusable waste which commands no market price and have to be dumped or discharged into the environment. Thus some wastes may be termed 'goods' and others 'bads'. Over time, however, there is some ambiguity as to what constitutes a good and what a bad since the same community may treat wastes as a good at one level of output or point of time but bad at another.

Several factors impinge on these changes in preferences; first, society may have a satiation level of recyclable wastes or conversely, due to the existence of high fixed

---

* Some of this material has already appeared as Number 16 in the series of Discussion Papers in Industrial Economics, published by the Department of Industrial Economics, University of Nottingham.

costs, there may be some output of wastes which is regarded as a minimum for society to be able to technically or economically recycle and, in this case, too little waste may be undesirable. Second, there may be changes in the stock of information affecting either the technology of waste recycling or our knowledge of the impact some wastes may have on us and the ecosystem. Thirdly, we may have under- or over-estimated the quantity of wastes that are emitted from some process, merely because of imperfect knowledge and deficient monitoring. Finally, the price of substitutes or complement products may alter, thus changing the demand for the waste.

Evidence of this ambiguous state of bads/goods definition is plentiful. Holes in the ground caused by quarrying and similar operations have recently become a marketable asset (4) and even spoil heaps can be used for landscaping and ballast in construction. Recently a dyeing firm under pressure from the local authority to reduce the pH of its effluent has managed to use an acidic by-product of another part of its dyeing process to neutralise its main dyeing effluent (5). Another dyeing firm in the East Midlands does not pay any effluent charges to the local authority and river board because its effluent has a beneficial flushing effect on the sewerage system (6). Waste paper became a valuable commodity after substantial increases in the price of raw materials, yet short-run demand soon became satiated because, in this instance, the paper manufacturers had not the necessary production capacity for recycling (7). The chipboard industry arose initially because of the inflated price of timber but eventually it was discovered that chipboard had some properties that were technically preferable to timber so the wastes of timber mills now seem to be quite unambiguously good. There are also cases where, below a minimum , wastes are bad but above that minimum, wastes are good; the technical and economic problems of recycling small quantities of lubricating oil waste from small factories and garages is one instance, whilst it has for some time been recognised that secondary zinc and lead smelters which emitted only small quantities of sulphur dioxide were in fact in a worse position than those emitting larger quantities because, in the latter case, the sulphur dioxide emission could be converted into sulphuric acid, but this would not be possible in small firms since the amount was too small (8).

Of course, society may not initially recognise the side effects of some waste products through lack of information of the chemical and biochemical processes and so, over time, outputs regarded as good or neutral may come to be perceived as bad. The dumping of biodegradeable woodpulp waste, noise in discotheques and the use of DDT as a pesticide are cases in point. Community preferences and perceptions are central to our interpretations of pollution; secondary lead refining has had to comply with very stringent regulations concerning lead emissions, but in secondary copper, where lead particulates are a by-product, the regulations have not been the same because public pressure has concentrated on the obvious sources to the extent that emissions of lead from other non-ferrous refining sources are now probably higher than from lead smelters themselves.

Tastes, technology and information are clearly the important factors affecting the amount of pollution a democratic society feels is desirable. These factors are clearly often in a state of flux since perception and pollution consciousness may continually change over time even if the state of pollution itself does not alter. Evidence on smoke control zones in Sheffield suggests that the number of complaints concerning airborne emissions from industrial sites increased with increases in control by the local authority even though industry had in fact greatly reduced total emissions [9]. Our own evidence in this study also supports this view. It is the intention here to combine these factors into a partial equilibrium model which investigates the interactions between the output of goods and bads, and their relationship to economic growth and information.

A priori we might expect that economic growth is positively associated with the absolute level of pollution and possibly also the rate of increase of pollution output [10]. We might also expect that regulation of pollution through charges and standards would limit pollution and that, as society becomes better informed about the harm pollution does, the democratic process will force an unregulated market into regulated behaviour.

The model. The approach developed here is a partial equilibrium one and assumes society chooses some ultimate commodity of satisfaction (the good life) which is made up of a combination of both goods and waste products; the latter invariably result from the production of goods through the use of optimal combinations of various resources. Our analysis can be generalised to an n product case but, for ease of exposition, we concentrate on the case of two products (good or bad).

We do not distinguish between public goods/bads and private goods/bads since we feel that a socially optimal level of output will result from actual market behaviour being regulated through the ballot box. This may be something of an heroic assumption, although of course economists have in any case suggested that bribes and compensations may allow the output of a bad to reach some socially optimal level, even when this is at variance with what the firm would like to produce in isolation [11]. It is a necessary assumption to the model since we intend using community indifference curves in goods/wastes space to help describe optimal production paths for society. In fact, this may not be too far from reality; the Alkali Inspectorate in the UK may be thought to implicitly admit that some pollution may be socially desirable when it adopts the "best practicable means" as its control stance [12]. This is also true of the local authorities whose controls vary substantially with the location of the hazard and who take account of such factors as employment in a particular area. To a certain extent, too, the self-regulation of industry or its positive responses to complaints before legal action is evidence that the use of community indifference curves are quite permissible to trace out society's optimal behaviour when unregulated behaviour is an allowable option . Perhaps a more serious objection would be the incorporation of waste recycling, which is an essentially technical property into the community choice vector. This, however, may be admissible if we consider that recycling is merely a process for producing a marketable product with a positive price. Pressure on the firm to recycle may come from the market place or, alternatively, it may be promoted by government through exhortation or increased controls on dumping, since the cost of recycling is a function of the costs involved in the recycling process and the costs of avoiding dumping and effluent charges.

Initially we assume that technology is given although, clearly, changes in technology may increase or reduce the output of goods and bads. We also assume two processes capable of producing these products which, when combined, form the ultimate commodity of satisfaction. Consider Fig. 1.2 as an illustration of the principle with two goods G and W where G is the object of production and W is a usable waste. (We deal with the case of unusable wastes later.) These goods may be produced by the alternative processes PA and PB. The maximum outputs for a given expenditure on inputs that can be produced by

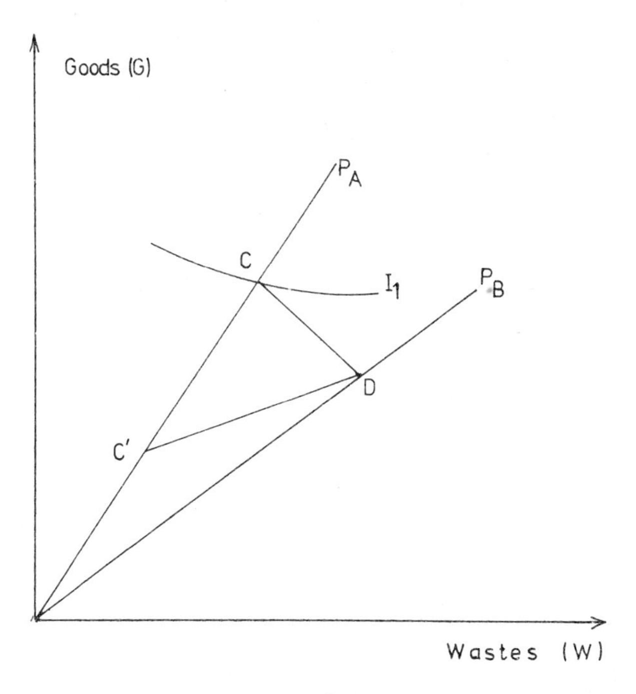

Fig. 1.2

PA is given by point C which represents given combinations of both products. The analogous quantities that can be produced with PB are given at point D. We do not feel we are violating the assumption of operating in commodity space even though ostensibly it might appear that we are dealing with a final and intermediate good. As Lancaster has pointed out, most goods of final consumption are merely inputs providing some further commodity of satisfaction.

The frontiers CD and C^1D represent the linear combinations of the two goods attainable with the two processes; we refer to them as the production possibility frontiers (PPF). In the case of the PPF with the positive slope (C^1D) has an absolute advantage in the production of both G and W. This model is, of course, only partial. When we consider the materials balance argument elucidated by R. Ayres and A. Kneese (13) which states that the weight of residuals generated must be equal to the physical level of output in any period allowing for embodiment in durables, etc., the partial model will clearly only be able to generate modest predictions relating to some particular situations on the impact of production on the environment.

There are clearly a number of different possible combinations of PPFs and community indifference curves which will yield entirely different solutions as to the means of production society will use. In some cases there may be only one possible solution for a given family of indifference curves and PPFs. This is so in those cases where there is a positively sloped PPF and a negatively sloped indifference curve (i.e. one process has the absolute advantage in producing both products and both products are goods) or a negatively sloped PPF and a positively sloped indifference curve (i.e. the case where no absolute advantage exists and one product is a bad which generates only disutility). The cases which are more interesting are those which involve changes in choice of process over the various ranges of economic growth and real income. This may occur when the slopes of the indifference curves and PPF have the same sign.

The slope and position of the PPF may change either through changes in knowledge of new technology or because changes in real incomes arising from economic growth allow more of both products to be purchased, thus moving the PPF outward. The shape and slope of the indifference curves are likely to change over income ranges since the income elasticity of demand for the waste (W) is likely to be less than that for the good (G), and indeed may become negative over some ranges. (Thus the idea of the throw-away pen, bottle, car, etc., originated in a rich country like the US.)

As our first case, let us assume that both the slope of the PPF and the indifference curves are negative as in Fig. 1.3. With the aid of this diagram we can show the simple case where, because wastes have a negative income elasticity of demand then economic growth leads to less waste output. In terms of the material balance argument, we cannot have less waste overall because of the positive relation that must exist between residuals and production. But at any one point in time we can have a situation as depicted in the diagram. Thus in Fig. 1.3, even though we begin with a corner solution at D with both a positively sloped efficiency frontier and indifference curve, concentration is removed from this process as the community's real income increases (which may happen as a result of both processes becoming more efficient or through an exogenous injection of real spending power).

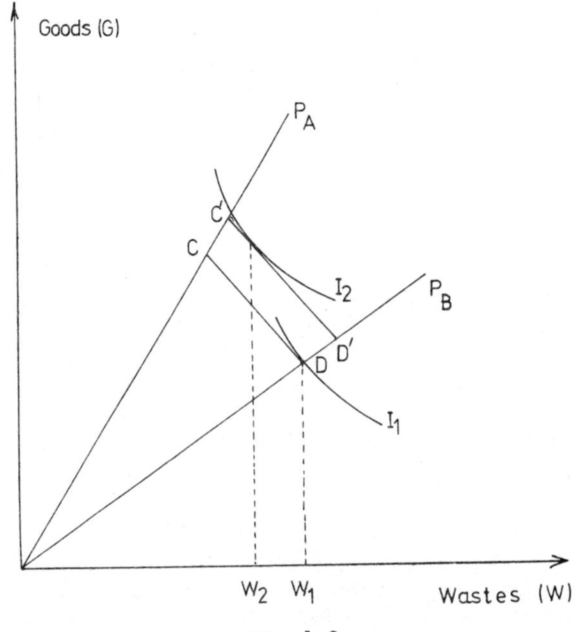

Fig. 1.3

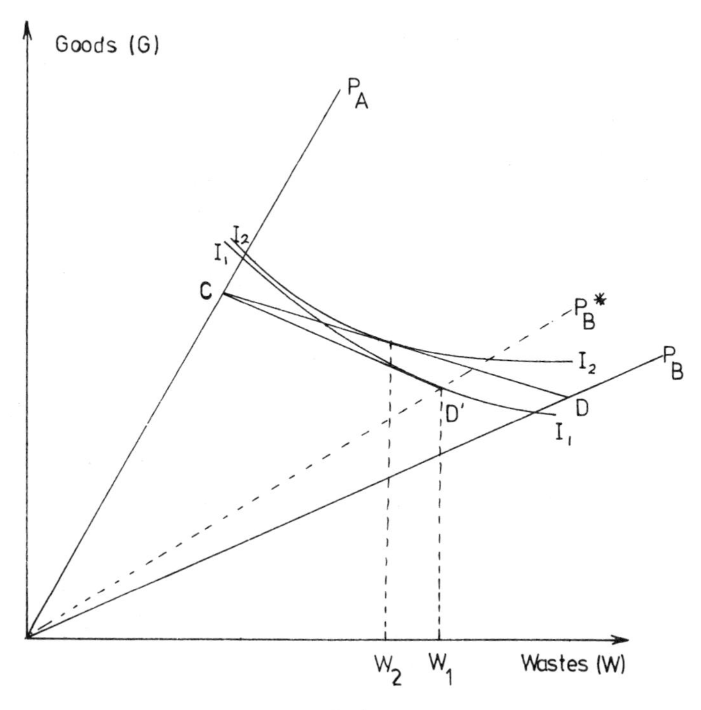

Fig. 1.4

A further development of this simple case is illustrated in Fig. 1.4. In this instance the community may perceive an increase in real income as a result of changes in information available. Thus in Fig. 1.4 the real efficiency frontier is CD. However, insufficient information on waste product extraction for process PB leads industry to perceive its efficiency frontier to be at CD^1. A priori it might be supposed that release of information would lead to process B being used more intensively, since the slope of the PPF has fallen. However, if good (W) were inferior then the income effect might outweigh the substitution effect and lead to a lower output of the waste product. Positive changes in technology would clearly lead to a similar result.

However, movement away from wastes may still occur even though wastes have a positive income elasticity. This is because the processes may be inferior even when the products of these processes are normal (14). If the income elasticity of good (W) relative to good (G) is low enough, PB will become an inferior process in the sense that its use will be inversely related to economic growth and real income growth, i.e. in this case with economic growth, less of the wastes intensive process will be used. We cannot, of course, conclude that economic growth will lead to less pollution in all areas of production but that it can in some. Over an infinite time horizon the materials balance principle will ensure that the output of residuals and growth are positively related overall. An extension of our basic diagram, as in Figs. 1.5 and 1.6 demonstrates this. Our analysis of inferior processes is aided by the construction of an unconstrained production curve (UPC). This arises through extending the PPFs of Fig. 1.4 to the axes and plotting the

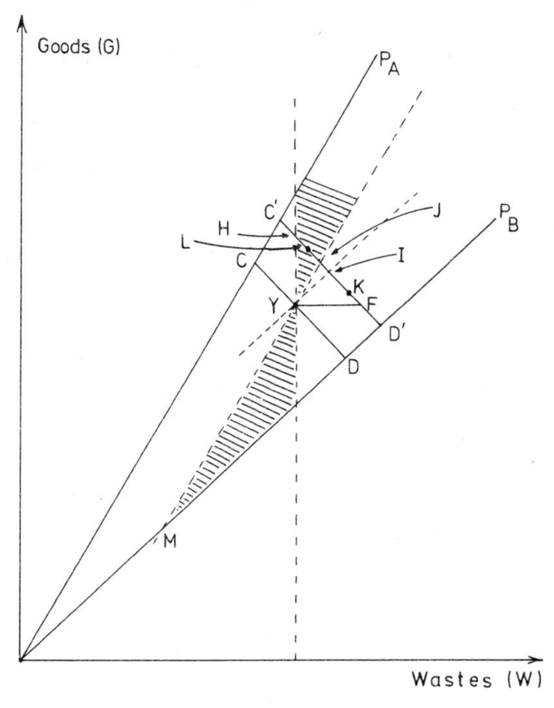

Fig. 1.5

points of tangency between the PPFs and the relevant community indifference curves, as in Fig. 1.6.

The extension of the PPFs to the axes is a natural result of real income changes that occur through economic growth or improvements in technology. We make no assumptions as to the distribution of technology changes between the two processes although intuitive logic suggests process A as the one which will receive the major share of technological improvement because of our expectations of tighter environmental control; if this were so, it would only serve to reinforce our prediction about the use of the waste intensive process and economic growth. If the UPC passes through point Y on Fig. 1.5, we can show the extent of usage of each process at that point by using a simple geometrical construction. We first draw a line parallel to PB through the same point. The extent of the usage of the various production processes is OM of PB and MY of PA. Now as the PPF moves outwards (through income or technical changes), to C^1D^1 for instance, we can draw lines through Y to cut C^1D^1 at H and F. If the goods are non-satiated (and thus non-inferior), tangency points between the PPF and indifference curves must lie on the segment HG. However, if the UPC cuts at any point other than the segment IJ, one of the processes is inferior over this range and increases in real income or technology will lead to a movement away from the inferior process. At I and J the income elasticity of demand for process PB and PA respectively is zero. Thus for PB the shaded area represents the situation where PB is inferior; this can occur if the income elasticity of demand for good (W) is low enough or that for good (G) is high enough. Thus as production of G and W rise, the use of PB will

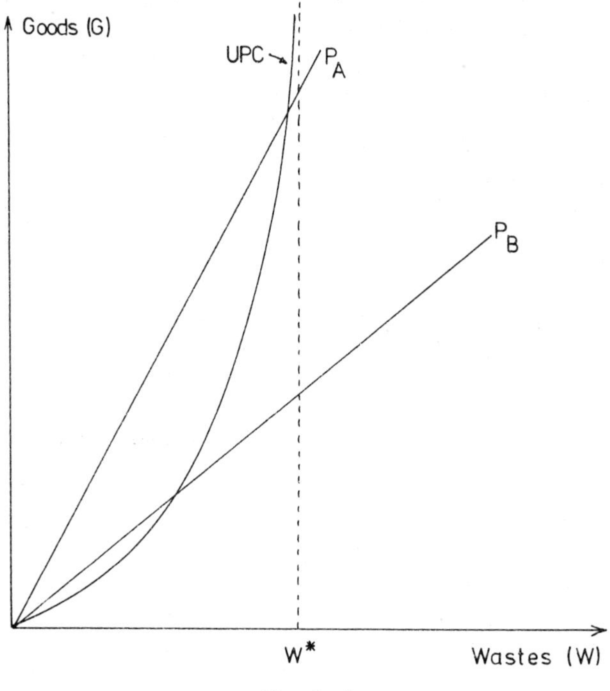

Fig. 1.6

fall absolutely; this can be verified quite simply using the graphic method by Dorfman [15]. The conclusion concerning economic growth and movement away from PB is reinforced if we assume that some of the goods are satiable (too much waste paper, for instance). In this instance we can construct a line W* as in Fig. 1.6 beyond which there will no longer be positive imputed prices for good (W). When the good is absolutely satiable it becomes a bad and along W* indifference curves cannot have a negative slope and along the UPC they cannot have a positive slope. Thus at some point the UPC cuts PA and the waste intensive process will be dropped altogether. Once again, our conclusion is that economic growth and pollution per unit of output (and eventually the absolute level of pollution) may be inversely related. This conclusion is not at variance with the views of economists like Kneese [16] who, for instance, wrote that ".... we can expect residual discharge to the environment to grow faster than national output .....", but it does suggest that residuals will be constantly moved to productive and least harm uses. Our analysis suggests that economic growth may lead to relatively more recyclable waste. Casual empiricism supports this view; there are few industries that, over time, have increased their emission levels per unit of output. Often the worst emissions come from new processes and in these cases increased emission levels are probably a function of society's imperfect information. With economic growth, society's collective view usually moves in favour of the least waste intensive process, as it has for ironfounding where there has been a rejection of the more efficient but more polluting hot blast furnace [17]. In aluminium smelting the pre-bake pot and the horizontal-spike Soderberg processes have replaced the vertical-spike Soderberg, not on efficiency grounds but because of existing and anticipated pollution control[18]. Spoil heaps per unit of coal produced are smaller and, in textiles, biodegradable dyes are now used almost exclusively. In many of these cases there has been a trade-off between output of goods, services and a cleaner environment although increasingly with industry anticipating controls or becoming more socially responsible, new research and development is automatically aimed at the least waste intensive processes. This may complicate the very important measure of the cost of pollution control since effectively we are dealing with an opportunity cost of new undiscovered innovations foregone.

The choice between goods and waste intensive processes seems to be typified by this general case of one normal and one inferior process but, as well as this general case, we may also have, and are indeed likely to have, Giffen processes in the sense that less of the process will be used as its relative price declines. This happens when society's income elasticity of demand for good (G) is so high compared to that for good (W) that it would be prepared to forego some W in order to get more G, after the relative price of W has declined. This can occur when we have non-satiated goods (W) and a downward sloping PPF.

In Fig. 1.7 the use of PB is positively related to the unit cost of production of G using PB alone. In the diagram a shift occurs from C^1D^1 to C^1D^{11}. Now if we assume the UPC passes through point Y and then draw a vertical line to cut C^1D^{11} and PB at I, the substitution effect must increase W and reduce G, the income effect increases both and equilibrium must be in the range HD^{11}. We now draw a dotted line through Y parallel to PA; if the UPC cuts C^1D^{11} in the segment HJ, PB will be used less as the unit cost of G (via its use) falls and vice versa. If the situation corresponds to this, the imposition of a tax on the waste intensive process will be self-defeating in the sense that it will encourage the use of the waste intensive process. The UPC is likely to produce this result at a level

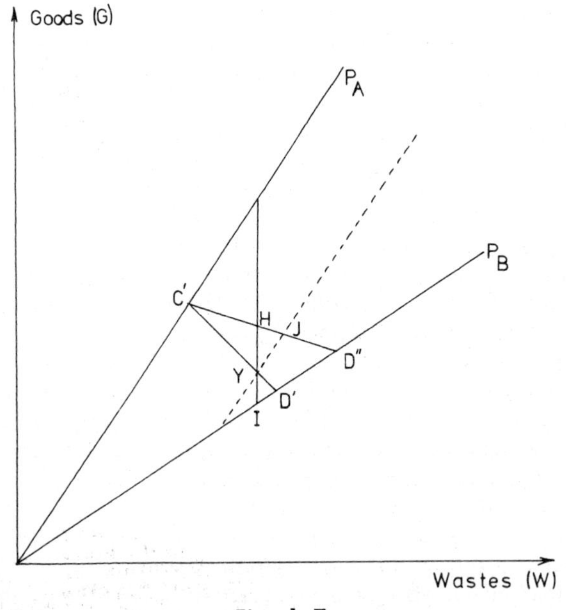

Fig. 1.7

W near the satiation level. The other interesting case is that where the waste intensive process has an absolute advantage in the supply of both G and W and where demand for W is satiated, PB is necessarily inferior and in this case the UPC will be leftward rising with increases in economic growth and good W will be Giffen unless the substitution effect is strong enough to overcome the income effect. Thus a process can be Giffen when one of the goods it produces approaches the range of satiation, even when that process has an absolute advantage in the normal and inferior good.

Our conclusions about economic growth and pollution still hold in the case where there is a minimum level of output W (which we refer to as W^{1*}) below which it is considered a bad and above which it is considered a good, although after some point it too may become satiated and revert to a bad again. Where PB has an absolute advantage, Fig. 1.8 illustrates the expansion path traced by the UPC. Where PB has no absolute advantage it would not be used at all.

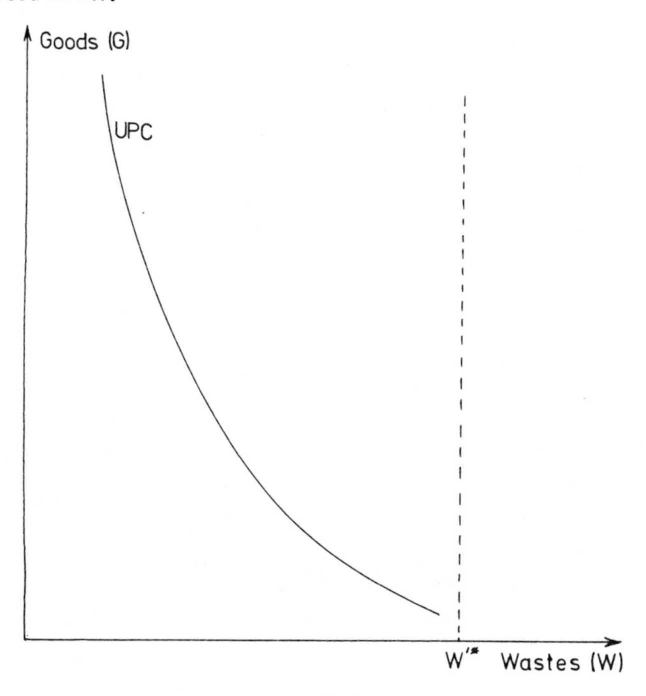

Fig. 1.8

Information and Wastes. So far we have assumed that society and producers have perfect information about the waste output of processes, recycling and the impact of the wastes on the ecosystem. Clearly this is likely to be something of an heroic assumption and below we examine the impact information changes may have on our analysis of goods and bads.

The problem of pollution (or bads) is that we may consume more of it than we want to, given perfect knowledge concerning the full extent of the bads produced. Thus in Fig. 1.9 society perceives the PPF as being CD whereas in actual fact it is CD^1; the full extent of bads emanating from PB has been underestimated. Toxic materials leaking into water courses, detergents causing long-term damage to the ecology of rivers, asbestos dust causing fatal illness as well as just respiratory difficulties to those inhaling it, and extended periods of damage caused by radioactive material may well be instances in point where, until recently, the full impact of these pollutants has not been recognised. In Fig. 1.9, PB* represents perception by society of the combination of goods and bads emanating from PB. Thus at point D society considers (wrongly) that production along PB yields W bads and G goods whereas in fact it yields W^{11} bads and G goods. Now, if society becomes more fully informed of the real quantities of bads from PB, several predictions may emerge. If at D^1 the slope I_2 is less than the slope of CD^1, after information has been received society continues to consume the same quantity of bads as before but has moved to a lower indifference curve and is thus arguably less happy than before.

Alternatively E may represent the new equilibrium point in which case consumption of bads has gone down from W^{11} to W^1 and, even though the point of equilibrium is on a lower indifference curve, it would be hard to argue that in real terms society was worse off. However, both cases show that potentially increases in information on the extent of pollution hazards may not theoretically be desirable. Thus we cannot always rely on a better informed society to necessarily control bads more than an ill-informed society, nor might more information always be better than less.

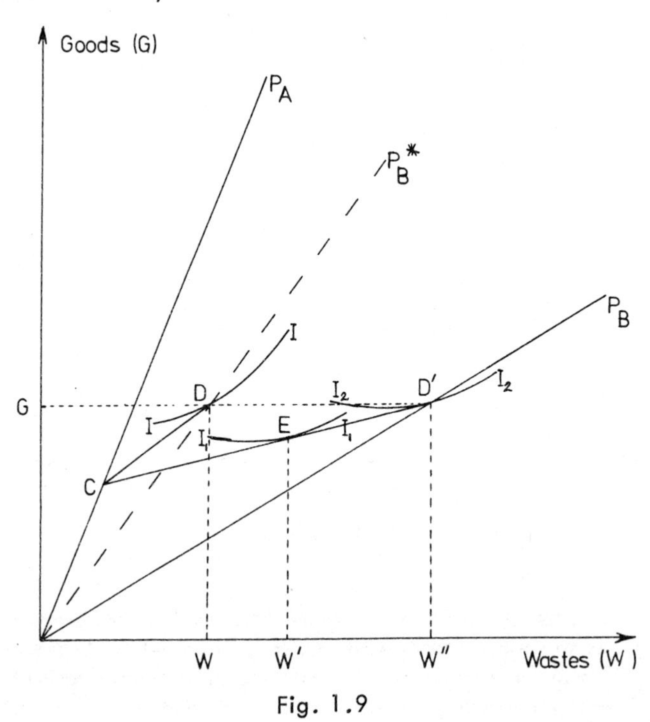

Fig. 1.9

So far we have considered the indifference curves to be independent of the production of goods and bads. However, this may ultimately be an incorrect assumption since the learning of tastes and knowledge of how 'bad' bads are and how 'good' goods are could well be a function of experience. Thus we would hypothesise that, as output of both goods and bads increases, the indifference curves would become steeper independently of inferiority of processes and differences in income elasticity for goods, thus increasing the likelihood of a C corner solution on PA and decreasing the possibility of the interesting theoretical result of an input of information failing to change consumption patterns but leading to a lower indifference curve.

It is also possible that information on the effects of bads (rather than their quantities) will also have some impact. Ultimately, the redefining of a bad into a good means that there is a change from a positive marginal rate of substitution to a negative one. In Figs. 1.10 and 1.11 two possible starting points are given which represent polar cases. Fig. 1.10 shows the necessary corner solution, concentrating on process A if there is no absolute advantage in the production of both G and W, and if indifference curves are sloped positively (i.e. the bad/good case).

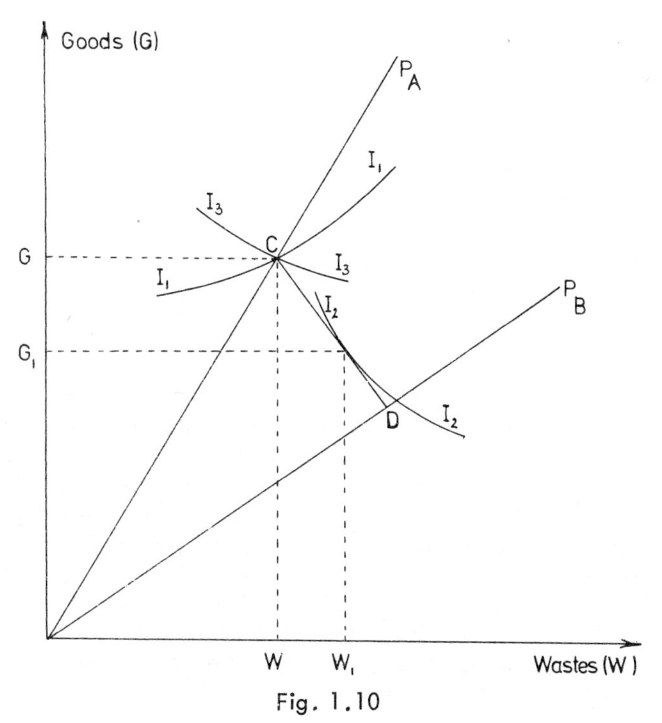

Fig. 1.10

Figure 1.11 shows a corner solution when PB has an absolute advantage in the production of both goods and indifference curves have a negative slope (the two good case). A change in information regarding the true nature of bads would normally result in some shift in emphasis from one process to another. In Figs. 1.10 and 1.11 society moves from W of wastes and G of goods to W^1G^1 of these goods following changes in information regarding the nature of the waste product. However, if the information changes redefined bads as relatively poor goods, in Fig. 1.10 (I_3 has a flat slope at C), or redefined goods as relatively "good" bads as in Fig. 1.11 (I_3 has a flat slope at D), then the corner solution yielding W and G of goods and wastes would remain unchanged. This "no change" situation would also be possible and likely if the efficiency frontiers had very steep

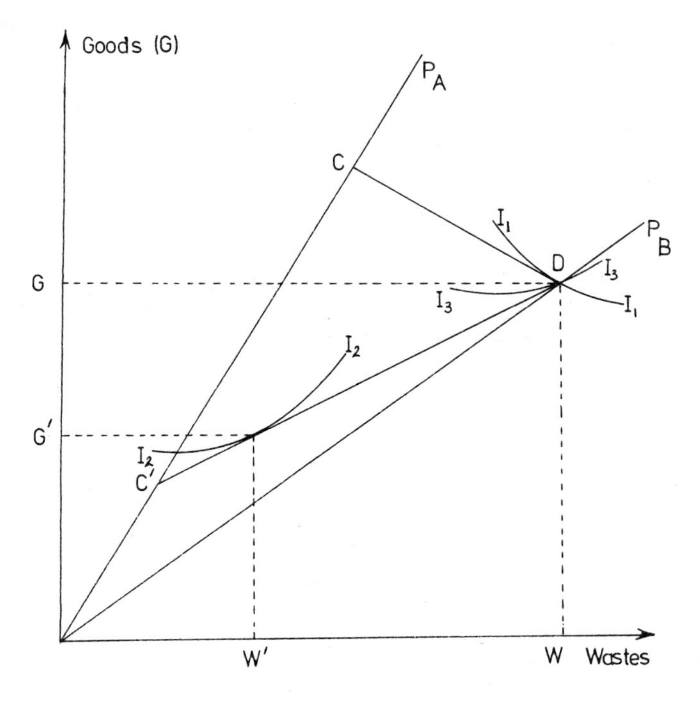

Fig. 1.11

slopes. That is, when the absolute advantage of PA in good W is very pronounced in the first case and where the absolute advantage of PB in both goods is relatively pronounced in the second case. Information regarding the suitability of waste for recycling and the true extent and nature of pollutants may be poor and this lack of information would be exacerbated in a situation where search was not worthwhile, as would be the case if there were no legal restrictions on effluent discharges or monopolistic buying of cheap raw materials by developed countries from underdeveloped ones. Changes in information about the potency of pollutants and possibilities for recycling may have a profound impact on choice but in other cases the only result will be that perceived welfare will fall and society continue to produce as it had previously.

Conclusion. We therefore conclude that it seems unlikely that, even in a market economy, pollution would go unregulated; the pressures of rising real income through economic growth are likely to lead to controls in some form or another as long as alternative processes both produce different amounts of goods and wastes and as long as preferences registered through the ballot box have some democratic outlet to the legislature. Residuals generation is related to pollution in any one of a number of complex interactions. However, economic growth might not in all cases increase residuals and therefore pollution. Generally we might expect that the provision of more information would also lead to less pollution and a welfare improvement, although there are cases where this might not occur. Systems of control usually chosen by government are either standards or charges, both systems of control may imply a reduction in the comparative advantage of the waste intensive process and both systems via the Giffen effect can have the opposite effect to the one intended without the goods themselves actually being inferior. However, on this count charges are much more likely to lead to a Giffen effect and so controls and restrictions may be a better option to take. Perhaps the major problem is the informational one; goods/bads definitions are frequently ambiguous and in a dynamic economic system new uses and definitions are constantly being found. The problem of control may ultimately be to relate current information to society's current preferences.

The Incidence of Pollution Controls on Firms and Industries

An important task for any policy maker who makes decisions when there are the competing economic goals of full employment, balance of payments, stable prices and a clean environment is to be able to predict the incidence of the taxes that are used to achieve one policy goal, maybe at the expense of another. The impact of pollution abatement on the firms and industries must begin with the effect on the firm's cost curve and the industry supply curve. It is outside our present scope to map out all possibilities but it will be quite clear that there are a number of possible permutations as to the incidence of the pollution abatement costs which will depend on the industry market structure, the motivation of the entrepreneur, the organisational structure of the firm, the efficiency of the flow of communication, the proximity of alternative processes, the possibility for the exploitation of large scale economies in pollution control and a whole range of factors governing the elasticity of supply and demand for the final product.

Consider firstly the impact of charges operating on the firm in the perfectly competitive industry. With no volume discounts or penalties, we might expect that under perfect competition the costs curves are as in Fig. 1.12.

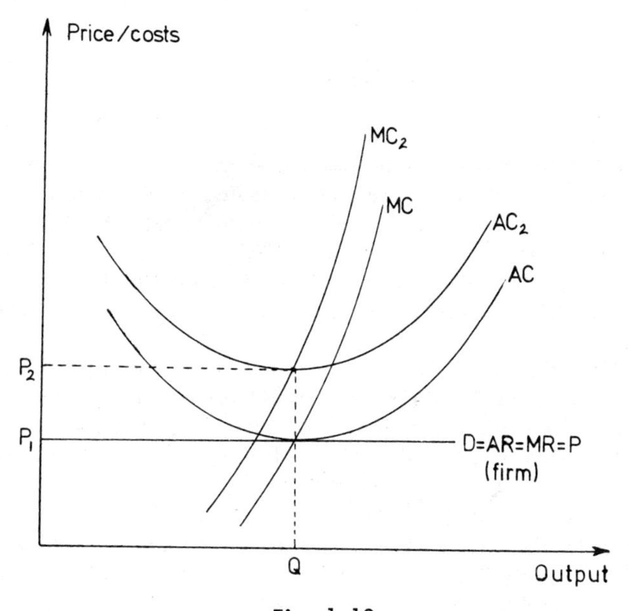

Fig. 1.12

In the short run prices must rise since firms will make losses at price P_1 and the total supplied by the industry must fall as in Fig. 1.13. The optimum size of firm, however, remains unchanged. Thus with charges the costs are passed on to the consumer in the form of higher prices but the firms have to bear some loss since because optimum output per firm remains the same, total quantity supplied by the industry is reduced and some firms must exit the industry in the face of these pollution abatement changes. The case of controls

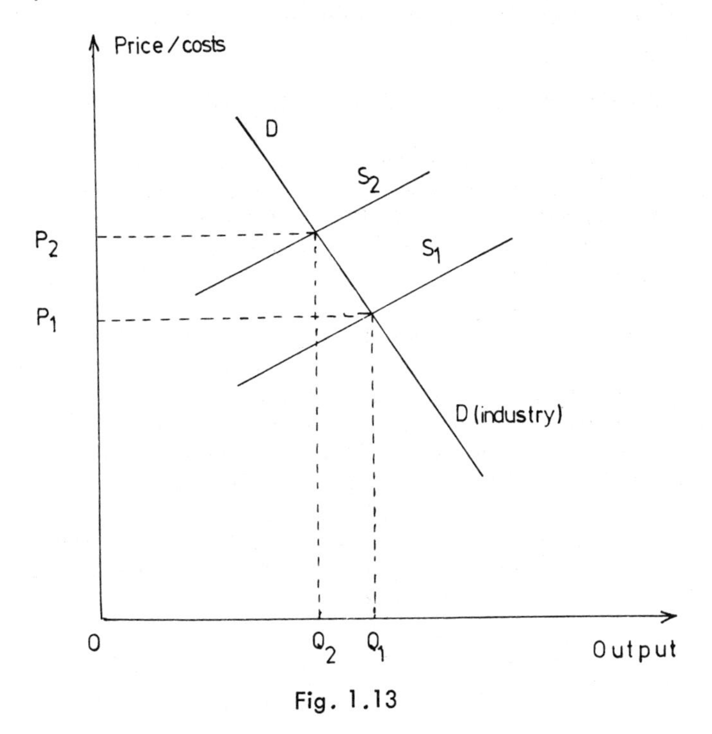

Fig. 1.13

which demand the installation of some pollution prevention device may be different. The differences will occur when the pollution control device is a fixed cost. What happens then is illustrated in Fig. 1.14. In the short run nothing happens, firms minimise losses and continue to produce at OQ_1. However, as before in Fig. 1.13, in the long term some firms have to leave the industry as prices must be raised in order to survive. Figure 1.14 shows that the new optimum size of firms is OQ_2 which is larger than before.

If we assume monopoly then we arrive at a slightly different set of predictions. In Fig. 1.15 the optimum output for the monopolist before he controls his pollution is OQ_1 at price OP_2. After the imposition of the charge the marginal cost curve will become MC_2 and the average cost curve AC_2. This will result in higher prices and reduced output (given by OP_2, OQ_2). However, if the effluent control takes the form of a fixed cost, nothing will happen since marginal costs will not be affected. All that we would predict under these conditions is that price and output would remain the same but profits may fall. In the long run, this may lead to some movement from one industry to another but this would depend on the varying rates of return between industries.

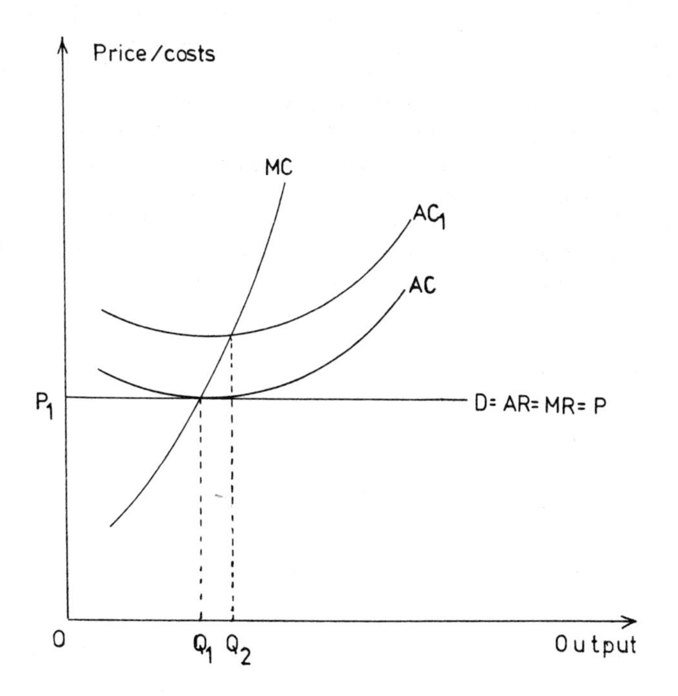

Fig. 1.14

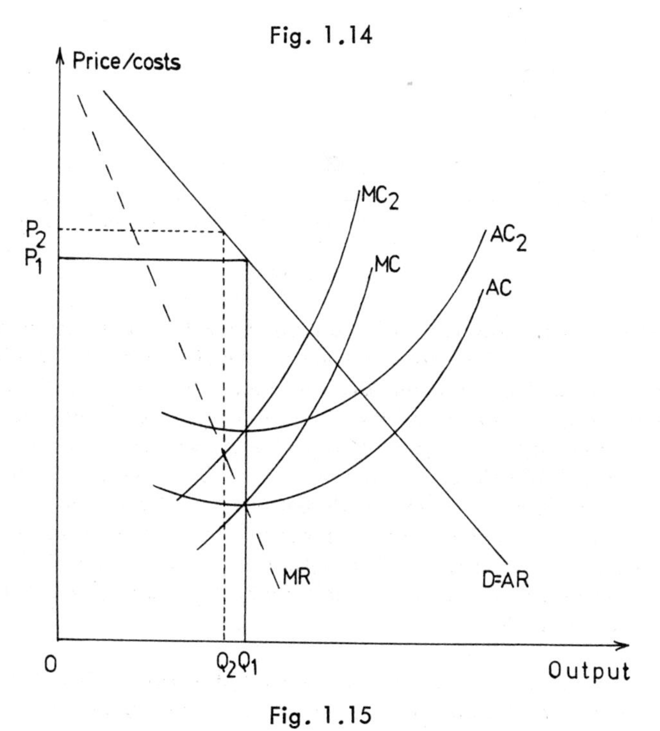

Fig. 1.15

Once we have used the perfect competition and monopoly models to illustrate the incidence problem, it is difficult to go much farther. Determinate oligopoly models are still rare and are not usually supported by adequate empirical evidence. The only other problems that should be aired are those concerning corporate motivation and organisation and the conditions governing the elasticity of demand and supply. If we use the sales revenue maximisation model subject to a minimum profit constraint as our guide, we are able to derive some useful predictions. Figure 1.16 illustrates firstly that the profit maximising firm earning pure profit will not change the level of output but will merely suffer reduced profits (point OX) whereas the sales revenue maximiser subject to a minimum profit constraint will first operate at OX_1 before the change and then at OX_2 after the change.

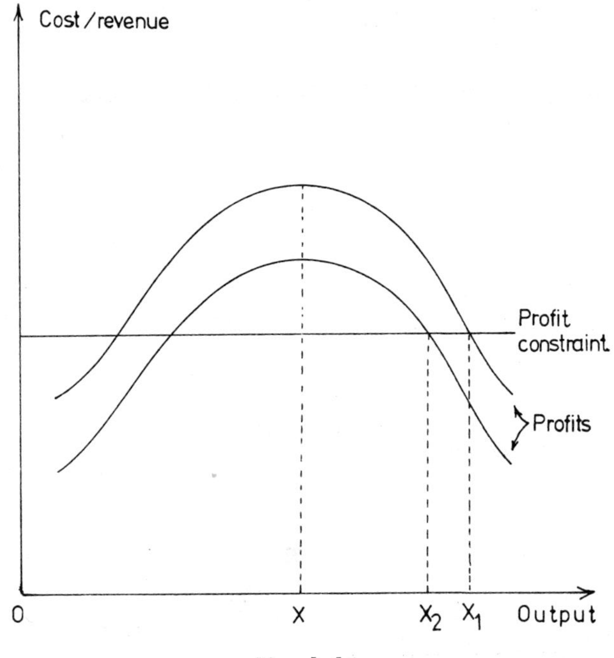

Fig. 1.16

Thus there will be a reduction of output and a consequent increase in prices. Behavioural theories of the firm based on ideas of organisational choice will predict little using aggregate measures of corporate welfare like profit and sales; the essence of their approach is that the incidence of pollution controls and taxes will depend on the impact they have on those corporate costs which determine prices and also under what level of slack, or inefficiency, the company operates. A number of empirical studies [19] have suggested that sometimes an increased exogeneous cost (like corporation tax, effluent charges, etc.) may set off a chain of events whereby the firm searches for new processes and products and eliminates some existing inefficiency. The result of charges may be to have a zero impact on costs because they act as a catalyst to reduce either inefficiency or pollution output, or both.

Finally, the impact of varying elasticity of supply and demand will affect primarily the incidence of the abatement costs in the corporate or consuming sector. Generally we can say that when demand is highly inelastic, then the consumer will bear the burden in higher prices whereas when supply is inelastic the burden will fall on the producer in the form of reduced profits. Figures 1.17 to 1.20 illustrate this. Tax raised equals P_2FHP and P_1 PHG is the portion of tax borne by the firm and P_2P_1GF is borne by the consumer.

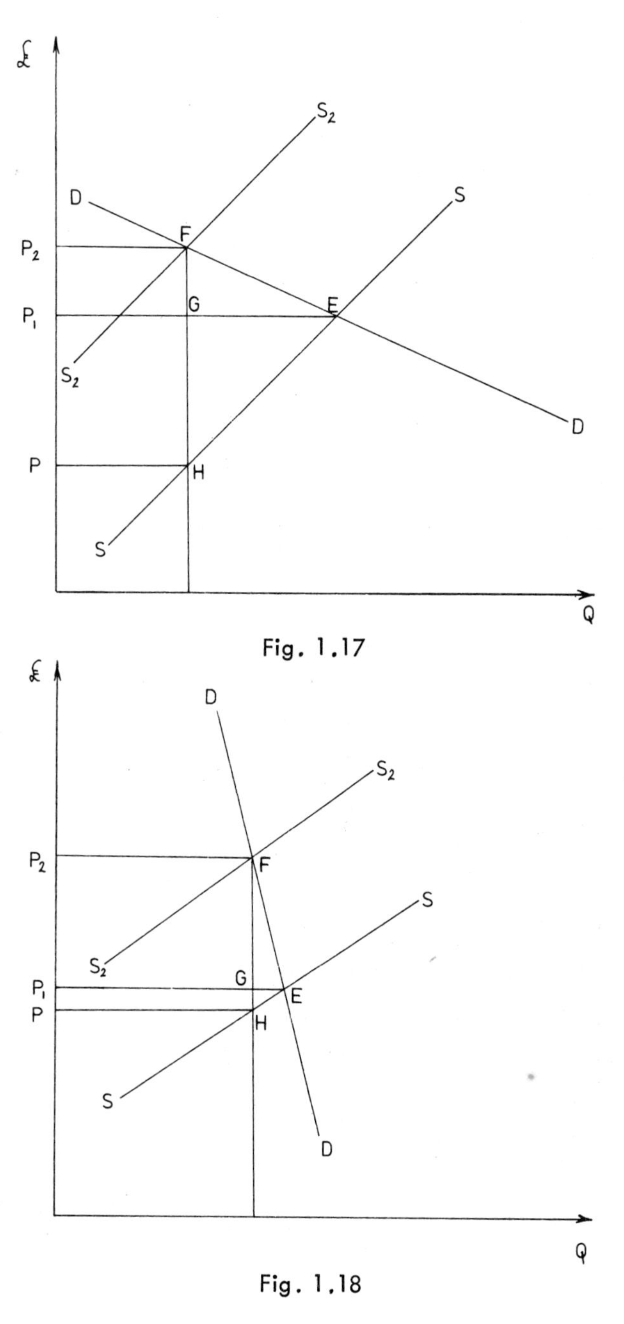

Fig. 1.17

Fig. 1.18

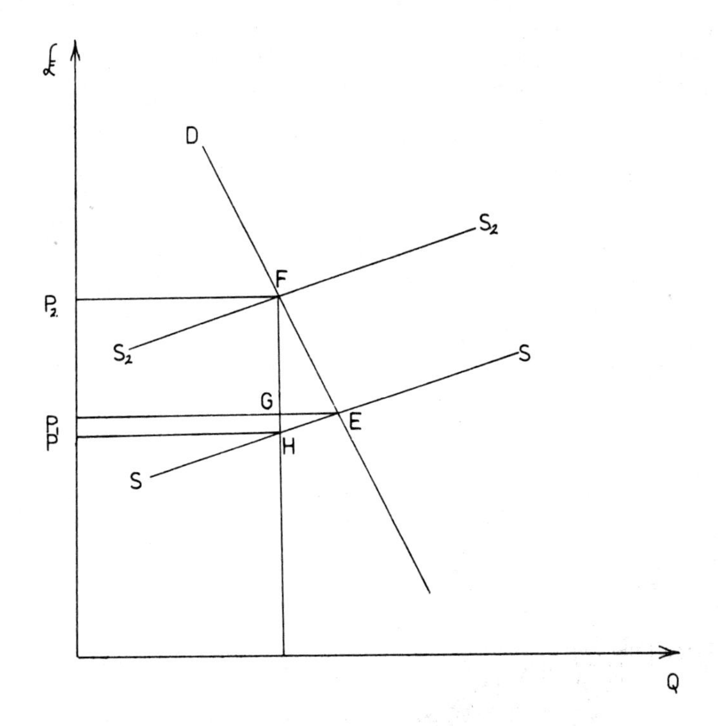

Fig. 1.19

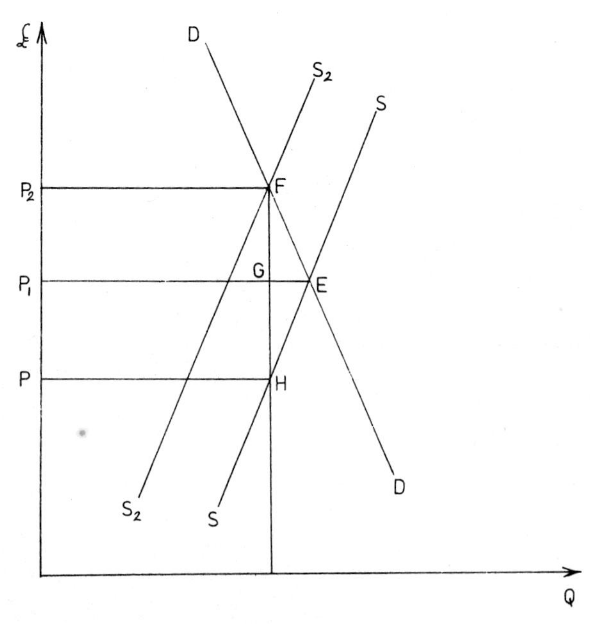

Fig. 1.20

Many factors will affect the elasticity of supply and demand but generally the larger the number of substitute products and strength of foreign competition will make supply elastic, whereas the number of substitutes is the vital factor affecting demand elasticity.

We conclude that adherence to the dictum that pollution is bad and must be stopped could clearly lead society to more loss than the loss it suffers from pollution. One factor that has to be taken into account is how a pollution prevention system is to be enforced and policed; the optimal level of pollution will depend on the enforcement and prevention systems and their incidence on consumers and firms. Whilst pollution cannot be claimed to be good, some may be clearly necessary.

So far, of course, we have omitted reference to the size of the firm and how it may affect the firm's response to pollution and its control. We therefore consider below the specific problems encountered by small and medium sized firms.

The Small and Medium Sized Firm

The importance of small firms in the economy has been the subject of considerable interest in recent years triggered by the Bolton Committee of Inquiry [20]. We do not wish to reiterate the Committee's findings here but should echo the feeling that focusing on the numerical importance of this type of firm often belies the real situation. They provide an important innovative stimulus to the economy and have a flexibility in the nature of outputs that larger firms are unable to supply, whilst they also provide an outlet for the natural desires for entrepreneurship.

There are also some factors operating on small firms both from the supply and demand sides which *a priori* are likely to be important from the viewpoint of pollution control. Perhaps the most important is the fact that small firms do not generally have the same access to financial and scientific resources that the larger firms do. This stems partly from their lack of size *per se* but also from their lower level of diversification. A large firm subject to sudden demands for capital for part of its operation may often be better able to meet these obligations through cross-subsidisation within the corporation; this is often not possible for the small firms. Raising finance from external sources is also likely to be more difficult for the smaller firm, either through ignorance of potential channels of finance or because their less diversified framework means that they represent a higher degree of risk to potential lenders. Lack of size and diversification also means that the division of labour, and thus the existence of specialised resources (particularly in the form of research and development departments), is more limited. The managing director may be production, research and financial specialists all rolled into one; keeping in touch with changes in technology and pollution control measures may frequently be outside the scope of the small firm entrepreneur. To a certain extent these deficiencies might be made up by the existence of research associations, even so *a priori* we might expect that impinging factors in the form of pollution control might present greater difficulties to the smaller firm which, whilst having smaller problems, may have fewer technical and financial resources for their solution by comparison with the larger firm.

Whilst the pollution impact of the small firm sector is on a much smaller scale and thus potentially less noticeable than that of the rest of industry, it might also be the case that the smaller firms may be politically less powerful at a national or local level. Their impact on employment and the total well-being of the country may be such that less account is taken of their preferences. Prior to the 1974 Control of Pollution Act this was noticeable to an extent when comparing the standards set by local authorities and the Alkali Inspectorate and even when comparing standards set within local authorities.

The problems of the small firm sector are likely to be exacerbated by the potential nature of pollution control measures. If, as we might expect, pollution control costs do not increase on a pro-rata basis with output and thus are open to the benefits of scale economies, it may be that better conditions within factories and better control of effluent from them might be linked in some way to size, although this would, of course, depend on the nature of the processes involved. To illustrate, the material costs of a simple cubic tank to hold effluent increases by the square, whereas its capacity increases by the cube. Scale economies should favour the large firm in other ways - from dust extraction equipment to complex metering and control of toxic wastes. Essentially control of pollution represents an increase in the complexity of technology and it is likely that the more complex technology becomes the greater will be the scope for division of labour; if this is so, the larger firm will usually have an advantage since the division of labour and specialisation are linked to the size of the eventual output. From the supply side, the organisation and behavioural response of decision makers within small as opposed to medium sized firms may vary because of the ways in which responsibility is divided and how information flows through the system. This potentially different behavioural response is likely to have an impact on the recognition of pollutants within the firm, the way in which controls are met and the incidence of costs onthe labour force and owners of capital. On the demand side, three factors are likely to be important discriminators between the small and larger firms. Firstly, small firms may be unable to pass on their increased costs of pollution control in the same way as the larger firms purely because greater competition between themselves will reduce their monopoly influence over their customers. We might expect then that, in the absence of collective agreements, profits may fall in the small firm sector and these competitive pressures may result in mergers and closures. Secondly, the small firms may be dominated by a single customer. This is particularly so in the East Midlands textile industry where chain stores like Marks and Spencer may buy all of a firm's output. This has essentially a dual impact of reducing the diversified nature of the firm's activities and the control the firm has over its manufacturing processes. Thirdly, the small firm that is relatively undiversified is more prone to cyclical movements in demand and if a trough in the business cycle should coincide with increased pollution controls then this could result in greater financial difficulties in the small firm sector. Consideration of the above factors means that with pollution control small firms are likely to present an interesting area of study.

References

1. A.V. Kneese, Environmental Pollution : Economics and Policy, American Economic Review Paper and Proceedings, 1973.

2. Control of Pollution Act, 1974.

3. P. Burrows, Pricing versus Regulation for Environmental Protection, in A. Culyer (ed.), York Economic Essays in Social Policy, Martin Robertson, 1974.

4. D.R. Davies and R.B. Levick, The Useful Conversion of Waste, The Royal Society of Arts Cantor Lectures, 1974.

5. J.S. Franklin and K.Barnes, Textile Effluent Treatment with Flue Gases, International Dyer and Textile Printer, September 1969.

6. The Knitted Textile Dyers Federation, Loughborough.

7. F. Livesey and V.A. Barcena, Waste Paper and the Market Mechanism, Long Range Planning, June 1975

8. Bureau of Mines, Control of Sulphur Dioxide Emissions in Copper, Lead and Zinc Smelting, Paper No. 8527, 1971.

9. G. Wall, Air Pollution in Sheffield, International Journal of Environmental Studies, Vol. 5, 1974.

10. A.V. Kneese, op.cit.

11. R. Coase, The Problem of Social Cost, Journal of Law and Economics, 1960.

12. Annual Report of the Alkali Inspectorate, HMSO, London.

13. R.V. Ayres and A. Kneese, Production Consumption and Externality, American Economic Review, 1969.

14. Here we generalise arguments of R. Lipsey and G. Rosenbluth on consumer demand, to the area of pollution, and production of goods and wastes. R.G. Lipsey and G. Rosenbluth, A Contribution to the New Theory of Demand, Canadian Journal of Economics, May, 1971.

15. R. Dorfman, Mathematical or Linear Programming : A Non-Mathematical Exposition, American Economic Review, 1953.

16. A.V. Kneese, op.cit.

17. The Report of the Grit and Dust Working Party, HMSO, London.

18. D. Rush, J. Russell and R. Iverson, Air Pollution Abatement on Primary Aluminium Potlines, Journal of Air Pollution Control, 1973.

19. R. Cyert and J.G. March, A Behavioural Theory of the Firm, Prentice-Hall, 1963.

20. Bolton Committee of Inquiry, 1971, Cmnd 4811, HMSO, London.

Chapter II

PREVIOUS RESEARCH ON POLLUTION CONTROL COSTS

Although pollution has been a matter of concern for many decades and much research effort has been put into studying its effects, surprisingly little has been contributed until recently upon the way in which attempts to attenuate pollution affect the economy and, in particular, the polluter. Thus much research effort has been aimed at investigating the impact of various levels of pollution on plants and animal life or assessing the impact of pollution on economic aspects of the environment such as the price of housing in a particular locality (1). The desire to impose pollution controls or to tighten existing controls further has resulted directly from this research and, whilst this is proper, very little has been done to assess the effects of these measures on firms, the individuals who work in them and on those who buy their products. This is possibly a direct result of adherence to the "polluter pays" principle; it is quite likely that many advocates of that principle firmly believe that if a corporate body is levied with an effluent disposal charge then it will be the corporate body alone which is affected. However, economists have for a long time been aware that the incidence of taxes on firms may fall on the customers and have effects on the size and structure of the industry. Thus the "polluter pays" principle really extends the concept of the polluter from the industry which supplies the goods to the consumer who demands pollution-ridden goods.

Industry Studies

Naturally there has been recently an increasing awareness of this problem, particularly in those industries in which pollution is especially acute. Thus in the UK the NEDO has studied the pollution problems of the Iron Founding Industry (2) and another study has looked at pollution problems of the Brewing Industry (3). Other studies such as (4) have considered the amount industry will need to spend on environmental improvement but in no way consider the impact these measures will have on those involved. Other similar work has been carried out by Trade organisations, although this is not generally available to the public; for example, the British Cast Iron Research Association (BCIRA) issued a report in 1972 on the impact of pollution control on the industry (5) and the Woollen Industry Research Association (6) has recently been conducting feasibility studies of water reclamation plants. However, little has been done in this country to assess, for example, price rises emanating from the abatement measures. Of course, as we have shown in the previous chapter, a great many factors must be taken into account in order for any reliable estimates to be made and in many instances the lack of data precludes the accurate computation of price changes.

Elsewhere the impact of pollution abatement measures has been more widely investigated. In the USA there have been studies of pollution abatement effects on the Meat Processing (7), Beet Sugar (8) and Secondary Aluminium industries (9) and other studies are described in a further report by the Environmental Protection Agency (10). In addition, two other studies (11 and 12) bring together the research onto a national basis to assess separately the impact of legislation for the purification of the air and water courses. Other work [13] by Leontief has incorporated pollution abatement in the national economy by means of input-output analysis.

International Agencies are also engaged in research along these lines; for example, the EEC is currently studying the position of the Iron and Steel Industry in the community and the GATT has been concerned with effects of differential pollution standards on international trade [14]. Other work on similar lines was reported more recently by Koo [15]. The OECD has also attempted to make an international comparison of the economic impacts of pollution control costs in (16), although the data used there must inevitably be regarded as incomplete.

It is difficult to generalise about the findings of these studies but one thing that does emerge seems to be the relatively small impact the necessary measures needed to contain pollution will have on the British economy - the OECD Report [16] suggests that the annualised costs of new programmes of pollution control * will be no more than one half of one percent of GNP in 1980, by comparison with one percent or more for those countries displaying comparable data. Investment in pollution control equipment as a proportion of total projected investment over control programme periods is less than half of that in other countries and, in Germany and the Netherlands, less than one quarter. From these data it would be easy to conclude that this country will come off lightly as the developed countries proceed to improve their environments but the figures must be treated with more than the usual degree of caution. There are two factors which determine whether the UK has a comparative (if not an absolute) advantage in pollution control. The first is that the natural environment's ability to deal with high levels of pollution depends to an extent on the prevailing weather conditions; one aspect of the English weather is that it is relatively good at dispersing or attenuating pollution (even if it does "export" the pollution to countries lying to the East). Directly connected with this, however, is the type of control system set up by the National Authorities. We can differentiate two basic types of control system - the determination of strict emission standards and what may be described as the "pragmatic" approach based on the ability of the environment to cope with pollution. There can be little doubt that a control system of the first type - since it ignores the disposal mechanisms nature provides free of charge - costs more to implement than one of the second type. In the past the UK control systems have been of this "pragmatic" type and the discretionary way in which they have been applied to some industries and not to others must inevitably lower the cost of pollution control. Those who would advocate a pollution control system based on output standards (which would not necessarily be without merit) would probably base their proposals on a claim that our control system led to more pollution than was desirable but whether the extra expense could be justified is problematical.

---

*It is important to stress that these data relate only to new programmes; they do not include existing pollution control measures. Furthermore, the definition of pollution control is by no means clear - whether, for example, car silencers or factory chimneys are regarded as p.c. measures is problematical. The only true indication whether or not the equipment used can be defined as pollution control equipment is if it has no impact on productive output.

Two further factors make international comparisons of control costs extremely tenuous. First it is difficult to assess whether the schemes that are being compared will end up with the same pollution levels in each country. From the environmentalist point of view, it would probably be sensible that they should, but an economist recognising different opportunity costs of pollution control in different countries would recognise that from an economic viewpoint each country should set different target levels of pollution dependent upon these opportunity costs. The economist would probably also recognise that there would be increasing marginal costs of abatement dependent upon the successes (or failures) already achieved and therefore the stage at which an individual country has already got to in its abatement programme is quite likely to determine future costs. Thus it may be that future air pollution abatement costs in the UK will be lower than in some other country that has not been operating smoke control schemes or their ilk. (Of course, the converse would be true if higher target levels of pollution control were set.) Secondly, it may be that factors such as the compact geographical nature of the country actually lower pollution abatement costs *. As will be seen later, it seems that fewer firms in the USA are connected to a municipal sewerage system and thus more will have to build their own effluent treatment plant in order to comply with the requirements. If there are economies of scale (and to a lesser extent economies of diversity) in water treatment (which we believe there are), control costs per unit of abatement must be lower in the UK than in the USA merely because of the proximity of municipal treatment plants.

Nevertheless, even if the proportion of GNP to be devoted to pollution control is only very small, the absolute magnitudes of the sums involved may be very large, indicating the enormity of the control schemes. Thus half of one percent of GNP represented approximately £300 million in 1974. A large proportion of this sum will probably go for new investment in pollution control equipment and we may thus anticipate a substantial increase in employment in this sector of industry. The extent to which employment in pollution-ridden industries will fall as a result of the control measures is difficult to assess but we cannot believe that no fall would take place.

The reason such an assessment is difficult at this stage is that, as we have already pointed out, so little research has been undertaken in the UK; the only really relevant work known to us is that by the Grit and Dust Working Party [17] and that by Bidwell [3]. The Grit and Dust Working Party provide data on the cost of changeover from wet arresters to high energy scrubbers. For example, a 20 ton per hour cupola run for 1000 hours per year would have total arrestment costs increased fivefold from £8000 with a wet arrester to £40,000 with high energy scrubbers (1970 prices). Further, these figures do not include capital charges and assume full capacity utilisation for the melting period - several factors can combine to double the costs. For a small plant a simple dry arrester will be sufficient but while such plant may be expected to have a life as long as that of the cupola itself and maintenance costs are limited to repainting and the occasional renewal of the baffle, the initial capital costs may run into four figures. For a cupola of 3 tons per hour melting capacity, a cost of £1200 was suggested to us and an additional £300 for a lined extension to raise the stack height to 65 feet. On the other hand, a wet arrester would have an

---

* The reverse could also be true. In countries where population density is high - as it is in the UK - there is less opportunity for the dispersion of pollution before it begins to affect a large proportion of the population.

anticipated life of only 7-10 years and have not inconsiderable operating costs. The capital costs (which may have doubled by now) would probably amount to £5000 for a 4 tons per hour plant and £8000 for a plant with 12 tons per hour capacity; in addition a settlement tank may be required which could cost £2000. Running costs would include the cost of electricity at around 2-5 kilowatts per hour and water charges of around ten pence per ton of metal; disposal costs could amount to several thousand pounds per annum.

There are alternative methods of pollution control; firms could install plant to recycle cupola gases by means of supplementary firing. Whilst this could cost £20,000 for the average cupola it would allow some reduction in fuel costs and could thus be amortised in two to three years. One firm in our sample had installed such equipment. Some firms were installing electric furnaces which, whilst they lead to a reduction in external pollution, may cause internal fume problems. Capital and operating costs of these furnaces are high; one firm suggested that capital costs are four times those of equivalent cupolas but the end-product is of a higher standard which probably has a lower price elasticity of demand.

As may be expected, there are substantial economies of scale in the operation of the high energy scrubbers; this is illustrated in Fig. 2.1 which shows the average costs (again excluding capital charges) of control for plants of different sizes. Such economies of scale are not apparent for users of the simple wet arresters; thus one may expect that the abatement standards will discriminate in favour of very small producers and very large producers ; the intermediate producers will have to install more expensive control plant and yet not be able to reap the potential economies of scale. There is some evidence that such polarisation is, in fact, accentuating a trend that may already be observed.

The essence of the Grit and Dust Working Party's Report is that the iron founding industry needed to spend an extra £60 million at 1974 prices during the period 1974-1982 following the introduction of new pollution abatement regulations. This sum represents approximately one third of the average investment in new productive plant and equipment. We would therefore expect the impact on an industry, already declining and with a profit margin on turnover of only 6% or so (2), to be substantial. The Government, recognising the seriousness of the situation, has allocated £50 million as special help to the industry (18).

Bidwell reports (inter alia) that one factor making this type of research difficult in this country is the "pragmatic" control system adopted. If output standards were clearly laid down in advance, it is argued, it would be comparatively easy to assess exactly what equipment will be required to achieve those standards and thus in turn the costings would be easier. But, with the standards effectively depending on local conditions, it becomes extremely difficult to apply any target pollution levels to the whole industry. We would concur with the argument although it leaves us in a schizophrenic position since, as economists, we would argue that the "pragmatic" control system which recognises the existence of free resources and which implicitly recognises inter-industry differences in control costs (even if measured on a cost per unit of abatement basis) should be capable of producing the economically optimal level of pollution.

Bidwell's thesis (3) attempts to assess the costs of pollution control in the Brewing Industry. After discussing five possible ways in which pollution may be abated * he proceeds

* A change of process technology or inputs; effluent treatment plant; changed methods of effluent disposal; "good housekeeping" and a change in the volume of production.

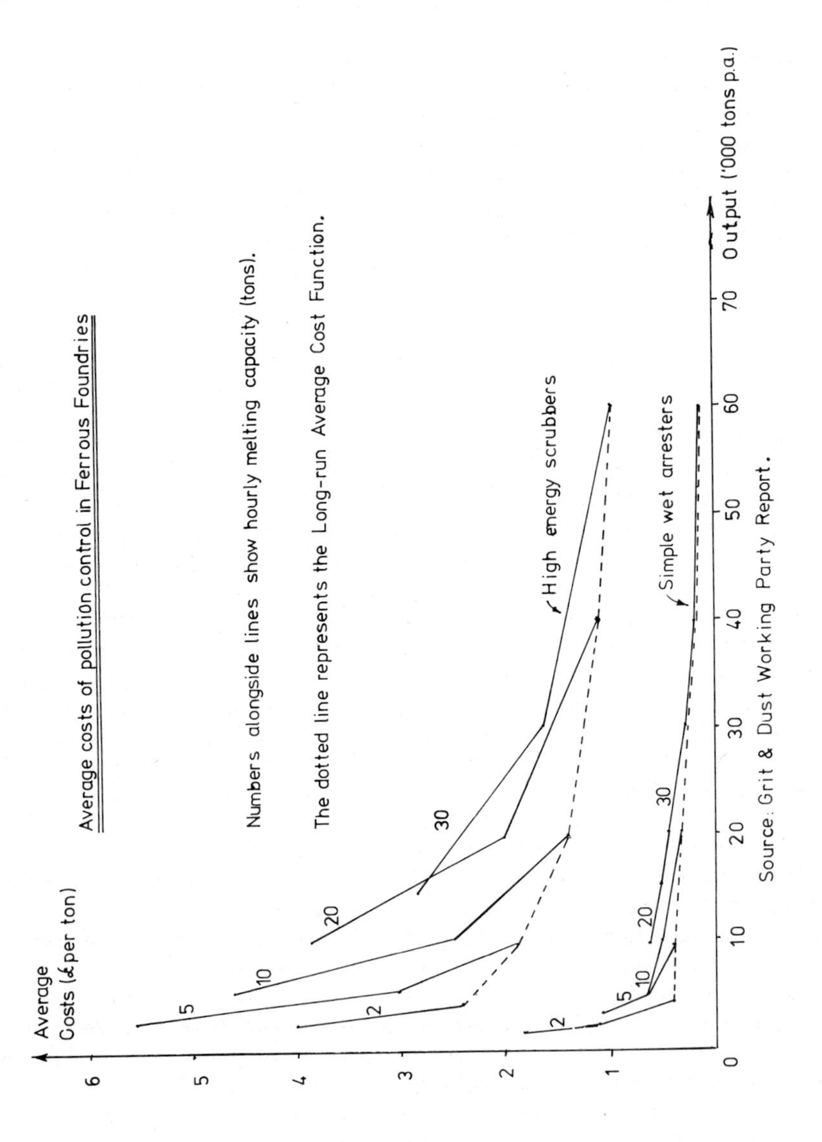
Average costs of pollution control in Ferrous Foundries
Numbers alongside lines show hourly melting capacity (tons).
The dotted line represents the Long-run Average Cost Function.
Average
Costs (£ per ton)
6
5
4
3
2
1
0
10
20
30
40
50
60
70
Output ('000 tons p.a.)
High energy scrubbers
Simple wet arresters
Source: Grit & Dust Working Party Report.

Fig. 2.1

to assess what effect improved standards of pollution control might have on future production costs. This is based on three possible future control standards and on data supplied by eight breweries accounting for nearly 20 percent of the output of the industry. Since there are five parameters identified by Bidwell as determining pollution levels, it seems unlikely that the claim that these eight are representative of the industry can be upheld. Nevertheless, using these data costs are assessed for some hypothetical (mythical) brewery which is somehow "typical". Although it was not clear on what factual basis the increased control costs for each level of abatement were obtained, it was concluded that only with the very highest level of pollution control would total production costs per barrel rise - and then only slightly. Less severe pollution control measures could possibly lead to a fall in total production costs on a per barrel basis as a result of the reduced water consumption consequent upon more efficient utilisation of resources. Further it was asserted that the present method of duty payment on beer operates as the highest disincentive to pollution above the costs of disposal of spent hops and lees, since it severely discourages wastage at the canning or bottling stage.

We cannot really accept this as the last word on the subject; there is no analysis of the effects of pollution control measures already imposed either on costs or on the structure of the industry and the data do not appear to us to be sufficiently good to be used as predictors for the whole industry, and there is no assessment of the implications of the necessary capital resources on the financial structure of the firms and the industry. Further if overall production costs really could be reduced by pollution control measures, it seems inconveivable that firms do not already operate such measures unless they are not operating in a profit-maximising environment.

Transatlantic research, however, seems to have been much more detailed although there may be two important factors that make such research easier in the USA than the UK. The first is one we have already pointed to - the control by standards system which makes industry-wide analyses much easier. The second is an apparently much more open approach to the use of financial data on firms by research bodies or governments. Here the 1948 Statistics of Trade Act prevents the disclosure of information pertaining to individual establishments and information collected by pollution control bodies is usually collected on the basis of confidentiality thus preventing public scrutiny or use. A comparison with the USA study on the Beet Sugar Industry [8] is revealing; there detailed information relating to all Beet Sugar Plants in the country is provided on a level that would seem unlikely in the UK.

This Study considers the impact effects of pollution controls on six main areas: prices, finance, production, employment, the community and balance of payments. The impact of standards * on prices was not expected to be large - roughly 40 percent of capacity in the industry already complied with the zero discharge requirement - and another 20 percent required a price change of only 0.2 percent in order to comply. Changes of such small magnitudes were thought unlikely to be passed on and, if so, were unlikely to have an effect on consumption patterns. Although the price changes may be very small, the financial effects were expected to be somewhat larger, particularly for smaller sized factories which currently had no pollution control equipment. Impacts on the larger

---

* As may be expected, standards relate principally to liquid effluent standards.

factories would be slight since there are clearly significant economies of scale in the treatment of effluent from these plants (note that only 5 out of 52 factories were connected to a municipal treatment plant and therefore almost all plants had to install some sort of treatment plant themselves). The smaller plants were thought likely to move from positions of small positive rates of return (of the order of less than one percent) to small losses (of the order of half of one percent) whilst the impact on the larger plants was small (profit rates falling from 6.5 percent to 6.3 percent in the extreme case); it was also felt that the smaller plants would suffer significantly by not being able to attract the new capital necessary to install pollution control equipment. These financial effects will obviously bear on possible plant closures and, on the basis of some fairly reasonable assumptions, it was thought that for four plants there was a high probability of closure; if more rigorous standards were to be applied, seventeen plants could face closure. The impact of these closures on the total level of production would, however, be slight in view of the new capacity under construction and employment effects would similarly be small; potential closures also mean potential losses to farmers if they cannot transfer their beet production to other processors and if this proved to be the case, losses could be considerable if the plants had been established through farm mortgages. Finally, the report concludes that balances of payments effects are not likely to be large in view of the new capacity under construction - little, if any, extra sugar would need to be imported as a result of plant closures.

The EPA Study of the Meat Processing Industry (7) reached similar conclusions; a higher proportion of these plants were connected to municipal treatment facilities than beet sugar processors and were therefore expected to experience no increase in costs due to the imposition of higher effluent standards. Those firms that would incur costs would have to raise prices by no more than 2 percent in order to maintain current profitability levels; in the event, it was thought that no price increases could be passed forward to consumers and some firms would therefore see lower levels of profitability. Profitability levels would be most affected in the canning industry - where it was already very low - with the smaller producers (with the exception of the very small which were connected to municipal facilities) faring worse than the larger producers. Projected closures were as follows :

	BPT	BAT
Small smoked meat plants	11	-
Small canning plants	5	-
Small sausage plants	-	12
Small mixed plants	-	11

(BPT Best pollution control technology available in July 1977)
(BAT Best available pollution contron economically achievable July 1983)

Although serious for the plants themselves, the closures envisaged would have little overall impact upon output of the industry - which operates in conditions of excess capacity in any case.

The EPA Study on Textiles covered a wider range of products than our East Midlands study, principally because it dealt with woollen goods. The study, however, omitted large integrated units because of their "much higher profits, larger size, integrated products, as well as their ability to apply a strong price on levels within the industry. The ability to raise (control) prices will allow them to pass on the cost of pollution control" (19). This assumption was largely borne out looking at the larger firms in their sample who came away from pollution control relatively unscathed. The methodology of the study involved splitting firms up into size/financial categories and then investigating the impact of control legislation on the groups given best practicable technology. To measure the impact of mill closings it was assumed that a mill would close only if the annual cost of treatment exceeded the annual cash flow of the mill. Within this it was further assumed that no mill would close if the estimated incremental operating costs were less than the minimum cash flow. When the operating costs of pollution control were greater than the minimum cash flow but less than the maximum, it was assumed a proportion of firms would close, the estimated number of plant closings was reckoned as a percentage proportional to the linear relationship between operating costs and cash flow. Thus if the cash flow ranged from $0 to $1000 and annual operating costs were $250 then 25% of all mills were expected to close. For the mills under investigation it was assumed that costs could not be readily passed on to their customers since the mills affected are the smaller independent companies which work in the market for the fill-in needs of the larger customers. It is these latter, the large corporations, who determine what prices to pay rather in the same way as the big multiple chain stores in the UK exercise power over their suppliers. Generally, though, costs could only be passed on at all when the market for the final product was strong. Within their sample there appeared very strong evidence that small firms were at a severe disadvantage when dealing with pollution control. Table 2.1 translates the original data from the study into minimum and maximum control costs as a percentage of selling price for small and medium sized firms in the industry.

TABLE 2.1 Treatment Costs as a Percentage of Selling Price

	Wool Scouring		Wool Dyeing		Woven Goods Dyeing		Knit Goods Dyeing		Stock & Yarn Dyeing	
	Small firm	Medium firm	Small firm	Medium firm	Small firm	Medium firm	Small firm	Medium firm	Small firm	Medium firm
Minimum treatment cost (if done via municipal sewage works)	1.5	0.12	1.4	0.7	0.4	0.2	3.0	2.0	3.3	2.0
Maximum treatment cost (best attainable means by the firm on site)	3.5	2.0	3.0	4.0	1.0	0.6	5.0	6.0	6.0	10.0

The variations are clearly substantial, particularly for those firms who have to do their own processing of waste; the absolute magnitudes, too, are in several instances large, especially for an industry that, according to the report, is not very profitable. These data led the authors to predict substantial plant closures and unemployment in the industry, particularly in the woollens section. Further it is predicted that, since many of the smaller plants dominate a local labour market, the community effects of these closures may be quite crucial. It was not thought that there would be much impact on international trade.

The report on the iron foundry industry [20] takes an even more pessimistic view and, incidentally, one which appears to heavily favour the larger firm. The report estimates that the cost of control will range from $1.37 to $14.05 per ton of castings depending on type of control and production output. Generally the costs of control declined geometrically with the melt rate as measured in tons per hour and often resulted in home producers being unable to compete with imports. This, among other factors, had led to an acceleration of shutdown of foundries, particularly those in the smaller size groups. Once again, as with the textile industry, whilst the national impact of the shutdowns may be minor, the impact on local labour markets was seen to be potentially severe.

These industry studies do not generally give any information relating costs of control to the degree of control achieved. Economic decision making requires data on average and marginal costs of control as well as total cost. The section below is concerned with a few studies which have attempted to deal with this issue.

Average and Marginal Costs of Pollution Control

Standardised data (quite unconnected with any specific industry) on the average operating costs (including capital charges) for eighteen different pieces of air pollution control equipment were published in an HMSO document Towards Cleaner Air [21]. These data, prepared by the Filtration Society in 1973, are based on equipment for cleaning hot dusty gases at the rate of 100,000 cubic metres per hour and, it appears, assume that the equipment is run almost continuously. The data also show the efficiency of each piece of equipment; if we were to consider a separate production process which we could call the production of clean air, an efficiency rate of 60 percent would yield an output of 60 units of clean air. The control efficiency may therefore be regarded as a direct proxy for output in the economists' terminology.

A graph of the average costs and control efficiencies shows that the relationship is strongly non-linear and regression analysis showed that the function form which gave the best fit was :

$y = ax^b$ where y represents average cost
and x represents control efficiency
and a and b were parameters to be found

The results of this regression were :

$$y = 9.9.10^{-7}.x^{51.8} \qquad R^2 = 0.77$$

Although not ideal - the function permits control efficiency to exceed 100 percent and hence an asymptotic function may be more realistic - this function does have the advantage of allowing marginal costs to be calculated easily. The average and marginal cost curve derived from these data are plotted over a range of control efficiencies in Fig. 2.2.

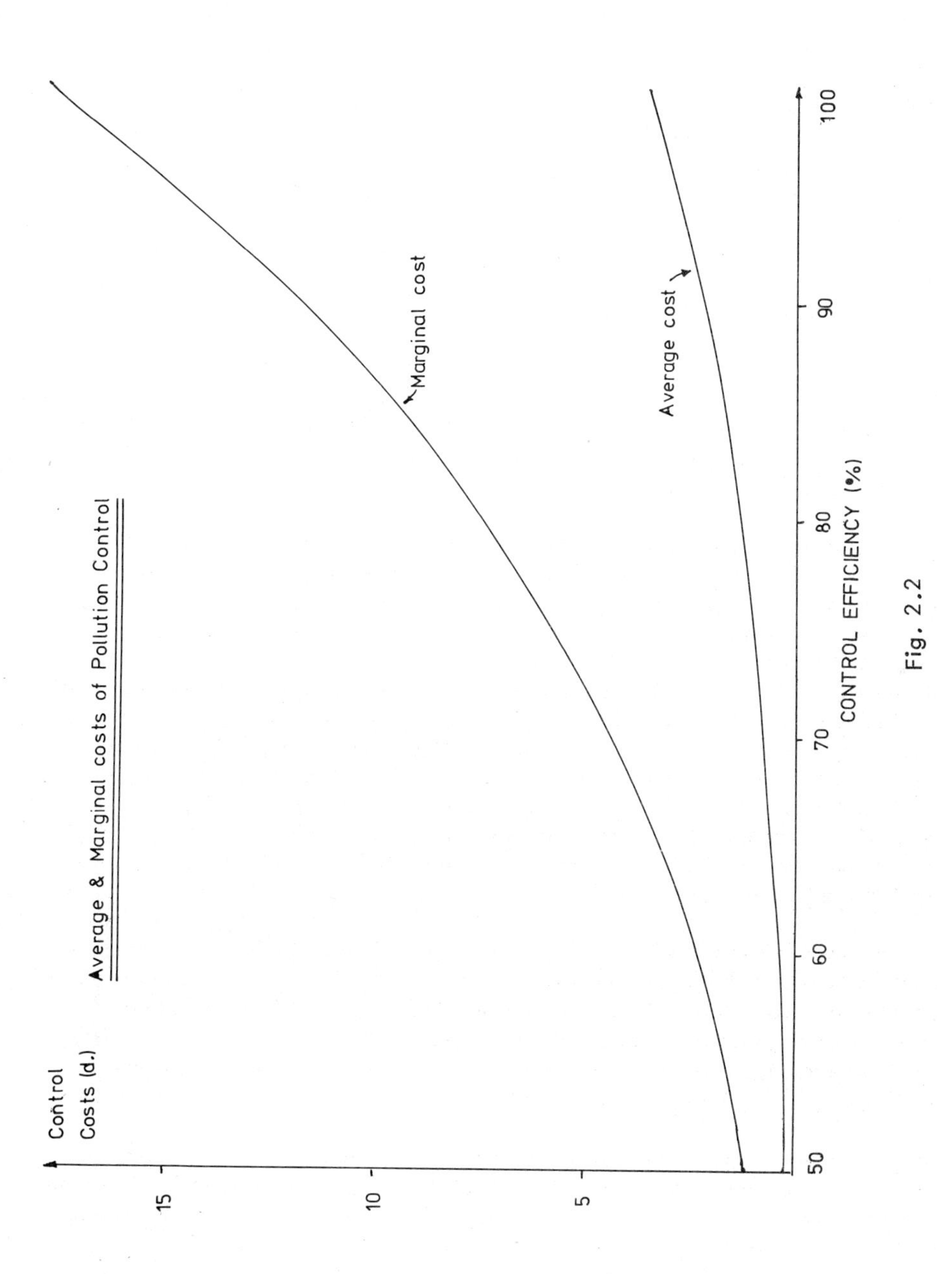
Control
Costs (d.)
Average & Marginal costs of Pollution Control
15
10
5
Marginal cost
Average cost
50
60
70
80
90
100
CONTROL EFFICIENCY (%)

Fig. 2.2

As can be seen from the graph, since average cost is rising, marginal cost is above average cost. It can also be seen that marginal costs rise at an increasing rate; this is a phenomenon one might expect as further purification becomes more and more difficult at high levels of efficiency.

The document referred to (21) discusses the choice of pollution control equipment from these data. Whilst we would not quibble with the decision process suggested, we must indicate that the matter of selection of the appropriate level of control is hardly discussed and is, in fact, assumed as a fait accompli. The graph would permit an assessment of the optimal level of pollution control; in the absence of other considerations this would be where marginal control costs were equated with marginal benefits from control. At the time of writing, we have no data on the latter but presume that the marginal benefit function will slope down from left to right and thus it is intuitively obvious that the "optimal" level of pollution would be somewhat less than 100%.

In the absence of such a marginal benefit function, it is still possible to use these data to assess the economic implications of control standards which may be laid down by regulatory bodies. For example a policy which demanded a 95% control efficiency implies a marginal benefit twice that implied by a policy requiring a control efficiency of 80%.

A similar study by Stone (22) gives more specific information on costs of pollution control in non-ferrous metal manufacture. The capital and operating costs, together with the respective control efficiencies, are given for twenty-two different items of dust control equipment. We do not wish to anticipate the results of further work to be undertaken by ourselves but it is already clear that these additional data show sharply rising marginal costs of control. We must also add that our new research shows that whilst an "average" control efficiency may be quoted for each piece of equipment, the marginal cost function should really be regarded as "quasi-discrete" in the sense that each piece of control equipment will only operate over a (fairly limited) range of efficiencies. We hope to expand on this aspect of control in a report we are now preparing on a detailed study we have conducted on pollution control costs in non-ferrous metallurgy and textile dyeing and finishing; meanwhile, we present results from a study of marginal costs of pollution control in the US steel industry (23).

The data shown in Table 2.2 relate to technology proposed by the US Environmental Protection Agency to help reduce emission from steel plants using electric arc furnaces. They refer to three systems - (a) a DEC system which is a direct furnace evacuation system connected to a baghouse, (b) canopy hooding which is an adjunct to a baghouse but which increases collection efficiency, and (c) a roof scavenger which would normally be attached to both (a) and (b) and basically helps reduce emission from the roof monitors.

A comparison of (a) and (b) first shows the substantial increase in cost to increase collection efficiency from 87.81% to 97.31%. A further comparison to (c) shows the incremental costs of reducing emission by a further 0.1% to 97.4%. Rapidly rising incremental costs are evident and their environmental impact, when viewed by the increase in efficiency from 97.3% to 97.4% is negligible, and may even be negative when account is taken of the pollution necessary to produce electricity to achieve this marginal increase.

TABLE 2.2 Total and Marginal Costs of Interlinked Pollution Control System

	DEC System (A)		Canopy Hoods (B)		Building Evacuation (C)	
	Total	Marginal	Total	Marginal	Total	Marginal
Efficiency %	87.8	-	97.3	9.5	97.4	0.1
Lb of dust collect/ton of steel	21.955	-	24.333	2.378	24.356	0.023
Capital cost - $m	2	-	5	3	7.06	2.06
Annual cost - $/lb dust	0.035	-	0.493	0.458	33.753	33.26
$/ton steel	0.77	-	1.778	1.008	2.543	0.765
Electricity required - Kwh/lb dust	0.116	-	7.786	7.67	689.786	682.00
Kwh/ton steel	2.552	-	20.79	18.24	36.47	15.68

This brief survey of published research shows that in not all cases is given the economic information necessary for public decisions and it does emphasise the considerable size and nature of pollution control costs in some specific instances.

References

1. J.H. Dales, Pollution, Property and Prices, Toronto, 1963.

2. Iron and Steel Castings, Industrial Review to 1977, National Economic Development Office.

3. R. O'N Bidwell, The Impact of Stricter Pollution Controls on Productivity in the Brewing Industry, Ph.D. Thesis, University of Bradford, 1975.

4. Water Quality 1973, Severn Trent River Authority.

5. BCIRA Confidential Report.

6. Financial Times, 2nd July, 1975.

7. Economic Analysis of Proposed Effluent Guidelines – Meat Processing Industry, US Environmental Protection Agency, Washington, July 1974.

8. Economic Analysis of Proposed Effluent Guidelines – Beet Sugar Industry, US Environmental Protection Agency, Washington, August 1973.

9. Economic Analysis of Proposed Effluent Guidelines – Secondary Aluminium, US Environmental Protection Agency, Washington.

10. The Economic Impact of Pollution Control, A Summary of Recent Studies, prepared for the Environmental Protection Agency, Washington, March 1971.

11. The Economics of Clean Air, US Environmental Protection Agency, Washington, March, 1971.

12. The Economics of Clean Water, US Environmental Protection Agency, Washington, 1972.

13. W. Leontief, Environmental Repercussions and the Economic Structure : An Input-Output Approach, Review of Economics and Statistics, August 1970.

14. Industrial Pollution Control and International Trade, GATT, July 1971.

15. A. Koo, Environmental Repercussions and Trade Theory, Review of Economics and Statistics, Vol. 56, May 1974.

16. Economic Implications of Pollution Control, OECD, 1974.

17. The Grit and Dust Working Party Report with special reference to cupolas 1974, DoE, London.

18. Trade and Industry, HMSO, 8th August, 1975.

19. Economic Analysis of Proposed Effluent Guidelines - Textiles Industry, US Environmental Protection Agency, Washington, 1974.

20. Economic Analysis of Proposed Effluent Guidelines - Gray Iron Foundries , US Environmental Protection Agency, Washington.

21. Towards Cleaner Air, HMSO, 1973.

22. E.H.F. Stone, Fume and Effluent Treatment Plant in the Non-Ferrous Metals Industry in Britain, International Metallurgical Review, No. 169.

23. J.E. Barber, Zero Visible Emission : Energy Requirements, Economics and Environmental Impact, Engineering Aspects of Pollution Control in the Metals Industry, Metals Society, 1975.

Chapter III

CHARACTERISTICS OF INDUSTRY STRUCTURE, POLLUTION AND THE SAMPLE

The Structure of Industry

We set out in Table 3.1 some regional data from the 1968 Census of Production. At that time nearly 7% of all manufacturing establishments in the UK were located in the region and, as can be seen, the region also accounted for the same proportion of manufacturing employment; other measures of the importance of the region (e.g. proportion of net output) also give rise to similar results. We therefore conclude that manufacturing establishments in the region are, on average, no larger or smaller than those in the rest of the country and the region may therefore be classed as "representative" in this respect. Column one

TABLE 3.1 The Importance of the East Midlands in UK Manufacturing Industry, 1968

	1	2	3
Food, drink and tobacco	5.8	10.1	8.0
Chemicals and allied industries	4.5	5.2	3.2
Metal manufacture	8.0	7.0	7.6
Engineering and electrical	8.3	24.7	21.7
Shipbuilding and marine engineering	0.2	2.5	0.1
Vehicles	7.0	10.0	9.6
Metal goods, n.e.c.	4.9	6.8	4.6
Textiles	16.6	8.5	19.2
Leather, leather goods and fur	8.9	0.6	0.7
Clothing and footwear	14.6	5.7	11.3
Bricks, pottery and glass	7.1	3.8	3.6
Timber, furniture	6.2	3.3	2.8
Paper, printing and publishing	4.4	7.5	4.4
Other manufacturing	5.5	4.2	3.2
All manufacturing	7.4	100.0	100.0

1 Proportion of UK industry employment in the East Midlands
2 Proportion of UK industry employment in the UK
3 Proportion of East Midlands employment in each industry

Source - Census of Production 1968, HMSO, Department of Trade and Industry.

of the table shows the proportion of each industry's employment that is located in the region; for example, nearly 6 percent of employment in the food, drink and tobacco trades is located in the region. It can be seen from these data that only the shipbuilding and marine engineering trades are substantially under-represented in the region (as may be expected, given its geographical nature); however, both textiles and clothing and footwear trades are over-represented. Nationally, there has been a substantial contraction of employment in the textile trades but because of the region's dependence on hosiery, which has not shared this national decline in employment, there has been little contraction in the East Midlands. As a consequence, the importance of textiles in the region is now even more marked than the table would suggest. The special importance of textiles and clothing is further emphasised by a comparison of columns two and three of the table. Column two shows the industrial breakdown of the national labour force and column three shows the industrial breakdown of the manufacturing labour force in the East Midlands; for example, nationally 10.1 percent of manufacturing employment is in the food, drink and tobacco industry but in the East Midlands only 8.0 percent work in this industry. Nationally only 14 percent or so of manufacturing employment is in both textiles and clothing and footwear, but over 30 percent of the region's employment is in these trades. Of course, the trades shown represent only broad categories and if we were to make finer breakdowns we would see the region being of even greater importance in specialised trades; for example, the importance of hosiery has already been stressed and further, nearly half of all employees in the footwear industry work in the East Midlands.

From these data we conclude that the nature of the trades in the region is unlikely to be too over- or under-representative of particular industries by comparison with the picture nationally and therefore the pollution problems encountered in the region may be expected to be fairly typical of the nation as a whole *. Thus we felt that an "across-the-board" approach to industry was the most appropriate and accordingly we drew our sample of firms in such a way that the importance of each trade in employment in the region was duly accounted for.

The Nature of the Sample

Initially a sample of 160 firms was drawn completely at random and not in such a way as to ensure a larger number of replies from any section of industry. In fact, we knew from the Census of Production that only a very small proportion of the firms would have more than 200 employees and we were therefore confident that most firms we contacted would be within the limits we had set.

We wrote to all firms asking them to complete and return a questionnaire (see Appendix 1). Inevitably, some firms had ceased trading since the sampling frame had been drawn up. In addition, some replied that they did not fall within the scope of the study since they were not manufacturers; we feel this is probably a slight mis-representation of the truth since these firms may have been manufacturers, albeit on a very small scale - butchers who make their own sausages are regarded as manufacturers for the purposes of the Census of Production. Follow-up enquiries were made of firms which did not reply within a given time limit and thus we were able to contact over 80 percent of our initial sample. Of this

* Both pollution and the location of industries tend to be highly localised phenomena. Thus although the region as a whole is fairly representative, particular parts (or sub-regions) are not so representative and these parts may therefore experience localised pollution problems as a result of localised concentration of particular trades.

number of firms, about 10 percent said they had no pollution problems and would not participate but 62 firms (nearly fifty percent of the remainder) did complete the questionnaire and gave us an interview.

The size distribution of these respondents is shown in Table 3.2 and it can be seen that the majority had less than 200 employees. A point that should be borne in mind, however, is that our enquiries related to establishments not enterprises and it is therefore possible that

TABLE 3.2 Size Distribution of Respondents in Initial Sample

Employees	Number
0 - 50	30
51 - 200	22
201 - 500	4
501 - 1000	6
	62

some of the respondents were part of a much larger organisation; indeed, we have direct evidence that this was true but that it in no way affected the autonomy of these firms. Given this and the fact that the number of medium sized firms is larger than we might have expected (although numerically quite small), we conclude that our sample tends to over-represent the medium sized firm. Nevertheless, we feel that this does not materially detract from the validity of our results. Furthermore, the evidence we have suggests that the size distribution of firms that did not reply was no different to that for those which did.

The industrial breakdown of the initial sixty-two respondents is shown in Table 3.3; as can be seen the largest representations come from the engineering and textiles, clothing and footwear trades. Naturally this is much in line with what would have been expected from the distribution of employment and we therefore feel that this sample is fairly representative of the region's industry. An analysis of the trades of firms that did not reply to our enquiry

TABLE 3.3 Industrial Classifications of the Respondents in Initial Sample

Engineering	16
Textiles, clothing and footwear	14
Metal manufacture	7
Plastic and leather	6
Timber and furniture	5
Food	4
Printing and paper	3
Building materials	2
Other	5
	62

shows that they too were fairly typical of the region's industry and we therefore conclude that our sample is not biassed either towards firms that had no pollution problems or towards firms that did have pollution problems and for whom our enquiry might have provided an opportunity to air any grievances that might have existed. It may be, of course, that some firms did not reply because they did not have any pollution problems (a few wrote and told us so) but we have no means of checking up on this.

It became apparent during the study that two trades faced particularly difficult pollution problems; these were the textile finishing trade and the iron founding industry. Textile finishers use very large volumes of water and consequently have potentially large liquid effluent disposal problems; as we have noted above the textile trade is an important employer in the region and we were not surprised to find that a quarter of all textile finishing establishments are located in the region.

The iron founding trade is also an important employer in several parts of the region and our evidence from the initial sample suggested that, because of the substantial air pollution problems both within and without the workplace, firms were faced with potentially severe technical and economic pressures. We therefore decided to look at these two trades in detail by contacting as many small and medium sized firms in these industries as was possible. Out of a total of 40 textile dyeing and finishing firms in the area that we contacted, 32 were visited, and out of 41 in iron founding we visited 29. This meant that overall we had achieved a response rate in excess of 80% in these trades and the cooperation we received was excellent. The size distribution of these firms in textile finishing conformed to that of industry in general but in the iron founding industry there was a distinct polarisation of firms into two separate groups of the relatively large and very small. Table 3.4 shows the national size distribution of iron foundries and the size distribution of the firms we contacted, and Table 3.5 shows similar information for dyers and finishers.

TABLE 3.4 Size Distribution of Foundries

	Sample %	National *%
Under 50	34	69.5
50 - 199	34	19.2
200 and over	32	11.4
	100	100.0

TABLE 3.5 Size Distribution of Dyers and Finishers

	Sample %	National *%
Under 50	12	59
50 - 199	66	31
200 and over	22	10
	100	100

* Source - 1968 Census of Production Order 313. Data exclude unsatisfactory returns from 100 establishments with average employment about 17 persons.

Air Pollution in the Region

In this section and the one which follows we do not wish to preclude the findings of the survey which are described in the following chapters, rather we seek to establish the overall position of the region *vis à vis* the national picture. The Warren Springs Laboratory has issued several reports on air pollution (1); these show that most air pollution (particularly in the form of sulphur dioxide emissions) is caused by domestic coal burning and not by industry. Naturally there are local exceptions to this but overall industry accounts for only about 20 percent of air pollution; the contribution from small and medium sized firms to air pollution in the region may therefore be expected to be quite small since they contribute nationally only 53% of the total net output of manufacturing industry and the East Midlands is generally representative of the national picture.

The air pollution levels of the East Midlands and neighbouring regions are shown in Table 3.5. Although not as bad as the North West and Yorkshire and Humberside, the region is worse than the West Midlands, which is often thought to have a much higher concentration of heavy industry.

TABLE 3.6 Average Winter Pollution at Town and Country Sites : A Table showing the Comparisons between the East Midlands and her Near Neighbours

	Smoke		Sulphur Dioxide	
	Town	Country	Town	Country
North West	144	83	179	125
Yorkshire and Humberside	125	74	168	100
East Midlands	109	52	140	69
West Midlands	79	41	129	58
East Anglia	68	25	114	-
South East	56	28	100	60

However, the fact that the sulphur dioxide to smoke concentration ratio is nearly one is indicative of a high incidence of domestic coal burning in the region*. In fact, domestic coal consumption per head in the region was, at the time of the Warren Springs Study, roughly three times that in the South East, as may be expected with a high incidence of coal mining in the area. Of course there are smoke-control areas in the region but in 1971 the Warren Springs Report concluded that "these are generally in the more lightly populated areas on the peripheries of the town or in the commercial town centre, while the more densely populated areas, the areas with the greater smoke pollution potential, are not covered". An examination of the maps of the smoke control areas leads us to conclude that, in the East Midlands, industrial areas are largely excluded with the possible exception of new industrial estates. This would seem to suggest that industry's air pollution problems are few and we were therefore quite surprised to find that this was not so. We must not paint too gloomy a picture of air pollution in the region. Pollution levels are falling

---

* This arises since coal produces most smoke against all fossil fuels.

steadily and this is likely to have been accounted for by the reduction in domestic fuel burning which has been of the order of 50% for the nation as a whole since 1960.

We have already mentioned the fact that individual factories may lead to significant localised pollution problems and the analysis above lends the impression that the region may be treated as one homogeneous unit; this is not so. The area north of a line drawn between Derby, Nottingham and Lincoln suffers substantially worse air pollution than the rest of the region, as evidenced by Table 3.7. We did not aim to specifically compare the problems over air pollution encountered by small and medium sized firms in these different parts of the region; rather we attempted to build up a picture of the region as a whole.

TABLE 3.7 Average Air Pollution at Town and Country Sites

		Smoke		Sulphur Dioxide	
		1968/9	1969/70	1968/9	1969/70
East Midlands	Town	119	99	144	136
	Country	54	49	66	72
Notts/Derbys	Town	131	111	147	141
	Country	55	50	67	75
Rest of Region	Town	92	76	138	128
	Country	49	44	57	61

Source : Warren Springs Report (op.cit.)

The geographic distribution of the respondents was fairly even throughout the region although, for specific trades, there was a much higher concentration in particular localities; for example, for historical reasons most dyers and finishers in Nottingham are located close to the river Leen.

Water Pollution in the Region

The problem of the pollution of our water system is serious, both from the aesthetic point of view and from the aspect of protection of our water supplies for industry and domestic consumption. According to a study (2) conducted by the Department of the Environment in 1970, on average eighty-three gallons per person per day of effluent were discharged. A further three hundred and twenty gallons per day were used as cooling water, mainly for electricity generation. Nationally, forty-seven percent of effluent was found to be of industrial origin so, unlike emissions into the air, industry is a very important polluter. The greater part of industrial effluent was discharged direct into the rivers or canals and fifty-six percent of all discharges (except in those cases where only cooling water was involved) were deemed unsatisfactory *. This amounted to fifty-four percent by volume

* A discharge is deemed unsatisfactory if it does not comply with the standard then required by the river authority. It does not imply any judgment of the standard and may, in fact, be questioned by the discharger.

of industrial effluent discharged. It is important to note, however, that total water used for cooling is more than ten times the usage of process water and that, even excluding electricity generation, process water usage is still less than cooling water usage. Data for a regional picture is confused because the Trent River Authority covered areas of the West as well as East Midlands but the evidence suggests that the East Midlands is cleaner than the rest.

In this area which covers most (but not all) of the East Midlands region, the percentage of industrial effluent is 43%. This is a little less than the national average, although use of water for cooling purposes is somewhat larger than the average with the massive concentration of electricity generation on the Trent which accounts for 92% of the total discharges in the region. Generally, the state of the water courses in the area is relatively good and the worst pollution occurs in the West Midlands, although some serious pollution is to be found in parts of the area (see Appendix 1 to reference (2)).

In manufacturing, the main discharges come from the chemical industry which is relatively widespread throughout the region. This and other major discharges are listed in Table 3.8.

TABLE 3.8 Main Discharges of Process Water (excluding electricity generation)

	Million gals/day
Chemical and allied	85
Iron and steel	40
Metal smelting and refining	27
Engineering	16
Paper	10
Food	9

Source : River Pollution Survey of England and Wales, 1970, pp. 39-46

Virtually all cooling water meets the river authorities' requirements; however, only seventy-two percent of process water was of a satisfactory standard, which is still in excess of the national average and the East Midlands area specifically is probably even better than this.

The expenditure needed on industrial discharges, although modest, is largely to improve discharges from iron and steel manufacturing; however, a large proportion of remedial works required at sewage plants may be accounted for by discharge of industrial effluent to the sewerage and this total expenditure required was forecast at £80 million. From Table 3.1 we have shown the major industries in the East Midlands were, in order of importance, engineering and electrical, textiles, clothing and footwear, vehicles, food, and metal manufacturing; these accounted for just under eighty percent of manufacturing employment in the region. Of these, only textiles comes within the first five national manufacturing dischargers of process water. To summarise therefore, water pollution problems are generated more by industry than by domestic users and the East Midlands has a relatively large discharge of liquid effluent. The standard of this is, however, substantially better than the national average and the worst water polluters among manufacturers, with the exception of the textile industry, are not highly represented in the area.

References

1. Warren Springs Laboratory Annual Report, 1971, HMSO.

2. Report of a River Pollution Survey of England and Wales, 1970, Volume 2, HMSO.

Chapter IV

ATMOSPHERIC POLLUTION

The Present Air Pollution Legislation

The intention of this section is to give a broad outline of the legal environment in which firms in England and Wales must operate; it is not intended to be comprehensive and for a more detailed discussion see Harris and Garner [1]. As has been already intimated, the following legal requirements apply only to England and Wales although there is similar legislation in Scotland.

When dealing with industrial air-borne emissions, it is convenient to divide industry into two parts: the so-called "scheduled processes" and others. This division arises from the operation of the 1906 Alkali Act through which the Health and Safety Executive (incorporating the former Alkali Inspectorate) controls the emissions of any processes which give rise to particularly dangerous or offensive emissions or those which are technically difficult to control [2]; included in these are chemicals, metal manufacturing and ceramics. None of the firms we contacted came under the aegis of this legislation although, in the past, one had done so. The main provisions of the Act are that :

i. no scheduled process may be carried on without a certificate of registration;

ii. it is a condition of first registration that the plant shall be equipped with the best practicable means of preventing the discharge into the atmosphere of noxious and offensive gases, (extended by the Clean Air Acts to include grit and dust) and for rendering them harmless and inoffensive where necessarily discharged;

iii. all apparatus must be kept in good order and continuous operation;

iv. registered works are subject to regular inspection; the owner must provide facilities for inspectors to enter and inspect the premises and carry out tests; he must disclose in confidence such details of the process that the inspector may require;

v. limits are prescribed for four processes only; for the remainder standards are laid down by the Inspectorate from time to time and adherence is accepted as evidence that the best practicable means are being used.

The 'best practicable means' is a subjective term that is intended to take into account the given state of technology, the cost of installation of the equipment and its running cost and the degree of harm done to the environment. Thus the required standards change over time and, although legal sanctions are available to ensure compliance, the Inspectorate relies upon cooperation and persuasion.

Control of emissions from the non-scheduled processes is achieved largely through the Public Health Act of 1936 (and as amended by the Public Health Recurring Nuisances Act 1969) and the Clean Air Acts of 1956 and 1968; these are enforced by local authorities and not by a national inspectorate as for the scheduled processes. All firms in our survey were subject to the provisions of these Acts. Dark smoke (that is, darker than shade $\frac{1}{2}$ on the Ringelmann chart) may only be emitted during strictly limited periods for the purpose of lighting up, etc. Further, all new furnaces must be capable as far as practicable of being operated continuously without emitting smoke when burning fuel of the type for which they were designed; a factory within an area which is subject to a smoke control order may not emit any smoke *. The Control of Pollution Act 1974 will further control sulphur discharge from furnaces (s.76) and prohibit the burning of insulation on cable (s.78) [3]. Local authorities also have responsibility for enforcing legislation requiring the best practicable means to be used to minimise grit and dust emissions from furnaces and to ensure that the emissions are within the specified limits for certain furnaces. They are also responsible for ensuring that the height of the chimneys for new or altered furnaces is sufficient to prevent the smoke, grit, dust and gases emitted becoming prejudicial to health or a nuisance. Another Act which is likely to have an impact on external air pollution (as opposed to conditions inside factories) is the Health and Safety at Work Act 1974 which, under s.3, requires all employers to ensure that persons not in their employment are not exposed to risks to their health or safety. It is too early to say what effect this very broad measure will have and it remains to be seen how active local authorities will be in applying this legislation.

Replies from the Random Sample

Air pollution from industrial establishments may be classified as arising from two distinct sources : (a) from stationary heating plant and (b) from the production processes themselves. Stationary heating plant may be used to warm offices and factories but it may also be used for steam-raising. In such circumstances, of course, the steam-raising plant is an essential part of the production process and thus it becomes difficult to distinguish the two sources. The distinction is important, however, since the nature of the pollutants emitted by the production processes is likely to be more varied and more harmful than the nature of those from the heating plant. Our interviews with firms lead us to the view that smoke control legislation has mesmerised some firms into believing that, providing their heating plant emits no smoke, they are in no way polluting the atmosphere if the production process itself emits no air pollutants. Of course, this is untrue; the combustion of fossil fuels leads to carbon monoxide and sulphur dioxide emissions, amongst others, which are invisible but potentially harmful to the environment.

We look first at the fuels used by firms for heating purposes. Since it was possible that the question regarding the division of fuels for 'heating' or 'power' purposes was open to misinterpretation in the case of firms utilising steam-raising plant, we have combined all replies to this question from such firms under the 'power' section where we feel that they properly belong.

There were comparatively few firms that used coal for heating purposes and, as can be seen from Table 4.1, the most common heating fuels were oil and gas. By looking at the

---

* Specific buildings may be exempted, however, on application by the firm.

annual consumption figures of their heating fuel inputs we are able to make an assessment of this aspect of industry's contribution to the pollution problem. One firm, employing between 500 and 1000 employees, used 30,000 gallons of oil per year solely for heating purposes. If we take the number of employees as 750, this figure represents 400 gallons per employee per year which compares favourably with an annual consumption level of roughly 800 gallons per year for the average 3-bedroom detached house. Considering that industrial heating plant is likely to be technically more efficient than domestic equipment, we are led to the conclusion that the impact of industrial heating plant on the degree of air pollution in the region is not likely to be great. Further, since the size of heating plant required may be expected to be directly related to the number of employees, this source of pollution is unlikely to be a major source of localised pollution except where numerous large firms are concentrated in a small area, although even here the use by these firms of much taller chimneys than domestic consumers will attenuate the localised air pollution problem.

TABLE 4.1 Usage of Fuels for Heating Purposes

Oil	Gas	Electricity	Coal	Total Number of Replies
34	14	12	12	46

Analysis of the geographical distribution of the firms using coal showed that most were located in North Nottinghamshire, although two were located in Leicester and the other in Derby. To a certain extent, of course, this reflects the geographical distribution of industry. These data represent only a 'snapshot' of the industrial scene at one point in time and many firms told us that they had at some stage been using coal but had changed to an alternative fuel. Although smoke control was sometimes an important consideration, we were left with the impression that the changeover had frequently been carried out on economic or technological grounds. For example, gas or oil boilers are generally much easier to control than coal-fired boilers and this may lead to lower running costs. Another consideration of industry may also be in terms of continuity of supply and storage costs which are both likely to be much lower in the case of gas or oil than in the case of coal. One factor leads us to believe that this source of industrial pollution will be further reduced in the future - the present Government's grants for improved insulation of factories will obviously lead to a reduction in the amount of fuel used.

We next turn to pollution from the production processes; Table 4.2 shows the replies obtained from the question whether the processes emitted air pollutants. Roughly 60 percent of the firms said that their processes did emit air pollutants. We have already

TABLE 4.2 Response to Question "Do your Processes Emit Air Pollutants?"

Yes	No
37	22

mentioned the difficulty in distinguishing the heating and process usage of fuels and here we must also question the way in which firms have interpreted the word "pollutant". It may be that some respondents have regarded any airborne matter - whether known to be toxic or not - as a pollutant or others may have regarded pollutants solely as airborne matter known to be harmful. The difficulties in perception of pollution are important; for example, some factories emit odours which, whilst harmless, are unpleasant and may be said to pollute the environment.

Table 4.3 below lists the types of air pollutants firms told us were emitted from their plant; Table 4.4 shows the different kinds of fuel usage for 'power' purposes. As can be seen, the most widely used fuel is, not unnaturally, electricity although fossil fuels are also widely used, and it is interesting to note the slightly fewer responses to the emission of carbon monoxide and sulphur oxides than there were firms using fossil fuels - this reinforces our earlier point that, to a certain extent, invisible emissions tended to be overlooked as pollutants. It was further reinforced on one visit in which we had initially contacted the company secretary for an appointment; whilst agreeing to an appointment he doubted whether the firm could be of much use since it 'had no pollution problems', but when interviewing the works manager it became apparent that the firm's heat treatment plant for metal emitted not inconsiderable volumes of polluting gases *.

TABLE 4.3 Nature of Air Pollutants Emitted

	Number
Carbon monoxide	12
Particles and dust	19
Sulphur oxides	14
Nitrogen oxides	8
Other	12

NB - Number sums to more than 37 due to multiple responses.

TABLE 4.4 Usage of Fuel for Power Purposes

Oil	Gas	Electricity	Coal	Total Number of Replies
15	10	54	7	57

If we discount these pollutants from the combustion process, the most important pollutant (in terms of the number of firms reporting) was the category 'particles and dust'. A wide variety of materials were transmitted in this way - wood and metal dusts were common but there were also instances of dust from machinery used to saw up cardboard, milk powder from a dairy and dust used in a printing works to prevent the off-set of ink. Some firms in the textile trades also had problems dealing with dust in the form of textile fibres. In some cases, the external air pollution problems caused by the dusts were fairly trivial -

* This also illustrates another point - the further the individual in industry is from the actual production process the lower is his perception of pollution problems. In terms of our in-depth interviews, this caused problems since we were concerned with collecting both information about the amounts of pollutants (best coming from the work's chemist or work's manager) and about the costs of control (best coming from the accountant or managing director). It was not always possible (except in the very smallest of firms) to get the best coverage in terms of the knowledge of respondents.

sawdust, for example, is probably not a long-term environmental hazard but in other cases this was not so. Although of numerical importance, we do feel, however, that the external environmental problems caused by industrial particulates are likely to be slight except in those areas where large numbers of factories exist side by side, although their impact on the environment within the factory is much greater. Nevertheless, some firms had installed equipment in order to prevent the emission of dusts and this equipment (filtration plant and so on) is not of negligible cost - the capital equipment cost for a modest engineering works could easily be of the order of £1000 or so and if the firm needed to install other plant, such as a cyclone collector for sawdust, the cost would be much more, although in this circumstance the firm might obtain some revenue from selling or otherwise recycling the dusts. Assessing the separate costs of internal and external control measures is left to the later section on costs.

As can be seen from Table 4.3, twelve firms reported that their production process gave rise to air pollutants other than the gases specified. Generally these pollutants were fumes of one type or another that arose from the production process; for example, one firm that was involved in a lot of soldering used hydrochloric acid for cleaning the copper and these acidic fumes were vented direct to the atmosphere. Some firms involved in metal casting used resin bonded sand for moulds which gave off offensive fumes when being formed and the metal poured. Similarly, firms involved in the plastics trade vented the fumes from the moulding process * direct to the atmosphere; occasionally things go wrong with these machines and some of the plastic burns to produce dense smoke and chlorine fumes. Whilst unpleasant, this was not a sufficiently frequent occurrence for the firms to have made any special provision for it.

Firms were asked what impact (if any) the imposition of the smoke control legislation had had on their plant or processes but few had made any changes. The legislation has, however, had a wider impact than this would lead one to expect since in a great many cases plant was introduced that satisfied the legislation before firms were required to do so by law. Very often this occurred when firms were replacing old plant when the opportunity was taken to meet likely future standards at that stage. Naturally, most local authorities give substantial advance notice of the imposition of a smoke control area and, in addition, some firms are able to meet requirements by applying for an exemption order permitting the emission of dark smoke only at strictly limited times. Although the smoke control legislation had little effect on most of the firms in our sample, for others the cost was not inconsiderable - a figure of 15% of new capital outlay being directly contributable to air pollution control would not be excessive or untypical. This could involve even very small companies in considerable capital expenditure. Whilst discussing the aspect of costs, it is appropriate to mention that often we felt that firms underestimated the direct costs of pollution control. For example, a tall chimney attenuates air pollution and in some instances firms were required by the local authorities to heighten their chimneys under threat of closure **. Even in these cases firms did not include chimneys in their pollution control

* We found no firms engaged in the manufacture of 'raw' plastics. Those firms we did find in the plastics trade were engaged in the moulding of plastic components in electrically heated moulding machines from pellets of plastic material. This is generally a very clean process since the plastic is not chemically transformed.

** In one extreme case a firm had been forced to move to a factory estate but, we are told, the local authority had stipulated maximum chimney heights for 'aesthetic' reasons although greater heights would have been technically more efficient.

costs although we feel they should. Other measures firms are required to make by law were also neglected when control costs were being reckoned. Whilst there may be benefits that have accrued to society as a result of these legal measures being imposed over a very long time period, it may be that a society starting out with no control measures would want to seriously consider the cost-effectiveness of all the measures it may take. Naturally, since one of our objectives was to measure the impact of pollution control costs on the behaviour of firms, we find it disturbing that most firms ignore the costs of some of the activities they are required by law to undertake. Indeed, this is all the more surprising since the firms which agreed to an interview may have been expected to be the more 'pollution-conscious' and possibly more than anxious to demonstrate how public-spirited they were and labour the point of how much it was costing them to be so. It may be, of course, that the measures taken so far have been those that are most cost effective and that, as marginal improvements to the environment become more and more costly to carry through, more and more firms and trade associations will call into question whether the industry can 'afford' these improvements.

In addition to the capital cost, most air pollution control equipment involves current operating expenditure. It is difficult to be specific here in view of the heterogeneity of the processes we were looking at and the nature of the control measures. However, some of these running costs are relatively easy to measure in terms of the differences in inputs - thus the biscuit factory which converted from coal to gas-fired ovens had negative additional running costs since the gas was cheaper than coal. In other cases, too, pollution control measures can have benefits other than simply a better environment. These additional benefits make it difficult to 'cost' the abatement legislation realistically - some businessmen would say that the new processes or technologies make commercial sense in their own right without the stimulus of legislation forcing the change upon them. Nevertheless, in other industries - typically those where technological progress is not rapid and where plant is relatively old - the cost of meeting air control legislation may be up to 15% of fixed capital.

A point of disquiet raised by some respondents was their perception that air pollution legislation was being enforced unevenly. We have sympathy with this view and are concerned since, if true, it can obviously lead to imbalances in economically efficient interregional and international flows of goods; however, we were unable to ascertain to what extent the complaints were more imagined than real. Of the bodies with responsibility for air pollution matters, the Alkali Inspectorate is vested with wider discretionary powers than the local authorities and, of course, some offices of the local authorities may be more strongly manned than others. Further, until smoke control zones cover the entire country, it will still be possible that some firms will remain discriminated against. However, recent developments mean that now there is only one statutory body responsible for air pollution and we look on this development favourably for it will remove any possibility of uneven application of the rules that could leave the worst industrial polluters (which we may equate with the large firms) largely untouched.

The Internal Environment

Many of the firms we visited employed some sort of equipment for maintaining the quality of the environment inside the factory. Table 4.5 shows the industrial classification of these firms; almost half of the firms without internal environmental control were in the textile and clothing trades and, in fact, most of these firms were small-scale makers-up of fabrics where we would expect few pollution problems in any case. There was no relationship between size of the firm and the use of internal control equipment.

TABLE 4.5 Industrial Distribution of Firms using Internal Environmental Control

	Number	Percent of Industry
Engineering	16	100
Textiles and clothing	4	29
Metal manufacture	7	100
Plastic and leather	5	84
Timber and furniture	3	60
Food	2	50
Printing and paper	1	33
Building materials	1	100
Other	3	60
	42	68

It is appropriate, however, to emphasise that many of the problems that necessitated this equipment were essentially of a very small scale and, in fact, some of the items mentioned by the firms were required by law. Of much more interest to us, therefore, were those cases where the problems were not on a small scale; we were able to identify 25 firms out of the total sample where we felt there were such large scale problems. Many factors may be expected to lead to these acute problems. Firstly and most evidently, it may be the nature of the processes being carried out in the factory; a key factor here is the size of the firm since larger firms may be expected to use a wider range of processes than smaller firms and thus the likelihood of encountering polluting processes increases in large firms. On the other hand, in the larger firm, the particularly nasty processes may well be hived off to specialist employees or located in specific areas of the plant; thus the number of people exposed to unfavourable conditions may be less but their degree of exposure much greater in the large firm than in the small firm. One factor that may enable firms to concentrate dirty production processes in designated areas of the factory will probably be the age of the factory or plant itself - we may expect a modern, purpose-built factory to confine dirty processes to particular sections, whereas this may not be possible where the factory is much older. We did not collect any information about the age of the factory but we did have information regarding the size of firms. After dividing the firms into those that had very small internal air pollution problems and those where we felt that the problems were more acute, we found that there was a clear link between the size of the firm and the size of the problem, as can be seen from Table 4.6 Nevertheless, over 60 percent of firms with large scale problems had less than 200 employees. For example, one firm in the metal manufacture industry and employing less than 200 people had spent £2000 in 1967 for equipment such as face masks and dust extractors to protect the men employed in the shotblasting section of the workshop. Just as the interviewer was leaving, however, the respondent remembered that the firm had spent an additional £35,000 on fume extractors, although earlier questioning had not revealed this. Further, two firms in the timber trade - both employing less than 200 men - had spent £14,000 between them on sawdust suppression equipment. Dust problems can also be acute in the textile industry since even small concentrations of lint in the atmosphere can cause byssonosis. One large

TABLE 4.6 Size Distribution of Firms Employing Internal Air Control Equipment

Size of Firm (employees)	Firms with Small Scale Problems	Firms with Large Scale Problems
Under 50	16	3
50-200	11	15
200-500	3	1
500+	2	6
	32	25

firm with over 800 employees said that, although only two or three operatives were involved, the capital costs of control equipment had been £5000 - admittedly a comparatively small fraction of total capital employed in the factory. Another large textile firm we contacted also mentioned this problem but indicated that it was not very important in the East Midlands since little cotton is processed there by comparison with the North West. The firm did indicate, however, that the employees had demanded 'muck money' for dealing with cotton materials and that the Trade Union had played a part in getting the firm to introduce control equipment. In other firms, some of this equipment is required by law, as in the case of grinding and fettling processes in the foundry trades. One large firm told us that approximately one sixth of new capital cost was accounted for by dust suppression equipment on machinery and that running costs were of the order of £1000 per annum.

Of course, these are all examples of cases in which the firm has been able to identify the problem and has done something about it. There may be many more instances in which the potential costs to industry are large if standards are to be improved. One chemist we spoke to in the textile industry said that he would be reluctant to work full-time on the shop floor since many of the chemicals in use were unproven. Sometimes it is not the chemicals themselves that are being used which may be harmful but combinations of factors can lead to the production of toxic substances - smoking in the presence of certain gases can be lethal but one Works Manager accepted rather reluctantly that he 'never could get that chap to stop smoking on the job' although the dangers had often been fully explained to him. A moot point here must surely be how far the company is prepared to go to sacrifice industrial disease for the sake of industrial peace. We spoke to several representatives of local trades unions and they told us that their unions were very active in this field and that their members were most concerned. Evidence from the firms, however, pointed in the other direction, suggesting that the innovations were primarily a result of management's intervention *, and secondly from the intervention of the Factory or Alkali Inspectorate. This evidence seems apparently self-contradictory but we feel that although trades unions say they are concerned about working conditions - and there is no doubt that pressure from employees can bring about changes on a very local basis- when it comes to national bargaining the unions do no more than pay lip service to working conditions since their primary (and traditional) objective is in terms of remuneration; this view was reinforced by a representative of one of the largest unions. We feel this situation to be unsatisfactory since our impression is that unions in other countries are not so restrictive in their outlook.

* One respondent said: "We are aware of the problems and we do it (sic) automatically."

Assessing the cost of internal air control equipment is not easy; we have already explained how easy it was for respondents to forget about important cost elements even when we might have expected them to have been the more cost conscious section of industry. The cost of providing a few gauze face masks is negligible for all firms but naturally there is the problem of getting employees to use them; this is a problem for management. However, since some of the processes (such as grinding and fettling) are required by law to have control incorporated, it is now impossible to buy machinery without this equipment and, without approaching the multitude of equipment manufacturers, it is impossible for a works manager to give any realistic estimate of what the machinery would cost without the pollution control equipment. Nevertheless we feel that a broadly based approach to the internal environment of the factory can be expensive; one firm mentioned the figure of one sixth of new capital cost and another indicated that equipment to control fumes cost three times annual profits. One firm involved in soldering processes kept the factory floor continually swept of solder droppings (which could be recovered) and lead level checks were a regular occurrence for all employees.

Despite these costs, some tangible benefits accrue to the firm. Firstly, a cleaner environment generally means a safer environment and this probably results in the loss of fewer working days through accident or illness. Indeed, work conducted by the British Cast Iron Research Association suggests that this is, in fact, true of founders. Further, better conditions may lead to lower labour turnover, reduced absenteeism and, possibly, lower monetary remuneration for the employees. Obviously the latter would not be a direct result of the cleaner environment but may arise because lower monetary inducements may be necessary to ensure the retention of staff. Many firms told us their main problem with staff was wages, but one trades union representative, speaking specifically about darkroom conditions, suggested that employers with the worst physical conditions paid more and had a higher staff turnover than those with the best physical conditions. Naturally this is a very specific example where the men involved are craftsmen; whether the same can be said for the less skilled trades remains to be seen. Thus we are unable to quantify these potential benefits at this stage except to say that, for some firms in some industries, the potential benefits may be large. This applies also to external measures to a limited extent; one firm we visited in the metal manufacturing business which had recently installed exhaust gas cleaning plant was relieved that workers no longer complained of excessive amounts of iron dust on their cars.

A Survey of Iron Founders in the East Midlands Region

We conducted a special study of ferrous foundries since substantial impact costs had been anticipated for new standards expected to be implemented in 1978. Forty-one firms were contacted, of whom one proved to be no longer an iron-founder, and of the remaining firms only eight refused to give us any information at all and twenty-nine were visited. This gave an overall response rate of 80 percent, which must be regarded as excellent coverage. In nearly half these firms it was the works manager we met, but there was a wide representation of other branches of top management. Perhaps the most unfortunate aspect was the relatively small number of accountants that we met; whilst many works managers had good ideas about capital costs of the plant within their control, fewer knew of operating costs and profitability levels.

TABLE 4.7 Geographical Distribution of Founders

Derbyshire	15
Leicestershire	4
Lincolnshire	1
Northamptonshire	4
Nottinghamshire	5
	29

TABLE 4.8 Nature of Location of Founders

Rural	3
Industrial estate	8
Urban	7
Heavy urban	11
	29

TABLE 4.9 Size and Organisational Structure of Founders

Number of employees	0-50	51-200	201-500	501-1000
Organisational type :				
Captive	1	4	3	4
Independent	9	6	2	0
	10	10	5	4

Tables 4.7, 4.8 and 4.9 give information relating to the sample of iron founders. Nearly half of the foundries contacted were in Derbyshire; the rest were more or less evenly spread around the other counties, with the exception of Lincolnshire. This meant that there was a slight tendency to under-represent foundries in Nottinghamshire but this arose since some had been covered in our earlier random survey. In view of the likelihood that foundries in locations of different types may well face different abatement costs, we show the nature of the location of the 29 foundries visited in Table 4.8. Naturally the distinction between these various types of location is fairly arbitrary and the definition of each one is somewhat subjective. Thus we have classified 'urban' sites as those where industry is located near to residential housing but not so surrounded that there is no room for expansion (this is 'heavy urban'). Further the 'industrial estate' category includes those firms that are located near other industrial firms and not those plants on specifically designated trading estates - only three foundries were located on such sites. Thus the order in which we have shown the geographical nature of these sites gives some impression of the potential need for abatement expenditure. It seems likely that, since nuisance clauses have formed the basis of previous control legislation, least pressure has been put upon those in the more rural areas for pollution abatement and, given the need for all firms to comply with the Grit and Dust requirements from 1978, firms located in this type of area will have to make bigger outlays on pollution control than ones in the more industrialised areas. On the other hand, over half the foundries visited were close to urban housing and one third so close that there was little apparent room for expansion of production on the existing site.

The need for abatement is a function not only of the location of the plant but also of its size and its melting method. To take the melting method first, Table 4.10 shows the melting methods adopted in the firms contacted.

TABLE 4.10 Melting Apparatus in Plants Visited

Cupola	- Hot blast	1
	- Cold blast	48
Oil-fired furnace		3
Electricity	- Arc melting	2
	- Mains frequency	8

NB - Figures sum to more than 29 due to plant of more than one type being operated on some sites.

The hot-blast cupola, whilst technically a very advanced type of melting equipment, has sufficiently acute pollution control problems that plants operating them come under the control of the Alkali Inspectorate rather than the local authority. In fact, the East Midlands region is not representative of hot-blast melting since, nationally, such plants account for 9 percent of melting capacity; nevertheless, although the region has many small-scale melters, the average size of firm is above the national average. As can be seen, the most common melting device was the cold-blast cupola and it is on these that attention must be focussed. (Oil-fired furnaces are generally very small and require little arrestment equipment, and electric furnaces generally create few external pollution problems except when oily scrap is used in them.) In addition to the plant shown in the Table above, there were ten induction furnaces used solely for purposes of holding molten metal. We therefore conclude that most firms in the industry operate plant of a type which is potentially subject to pollution control.

The size of the plant can be expressed in many ways; our definition of a "small firm" implies that we should measure it in terms of employment but, in view of the fact that the abatement regulations are phrased in terms of the hourly melting capacity of the plant, it seems sensible to use this measure rather than employment size, although it is likely that both would give similar pictures of the degree of concentration in the industry. The breakdown by melting capacity is given in Table 4.11; it can be seen that the sample taken is fairly representative of the industry as a whole, given the fact that the average size of foundries in the region is above the national average.

The types of arrestment plant recommended for future use are as follows: the simple dry arrester now in use is to be discontinued and replaced by the simple wet arrester for cupolas up to four tons per hour capacity; for cupolas exceeding four tons per hour and not exceeding ten tons, a multi-cyclone or low intensity scrubber is to be used and high intensity scrubbers are to be used on cupolas exceeding ten tons per hour (ten percent of all cupolas in use). The implication of these standards is that grit and dust emissions would not exceed 26 lbs per hour except for larger cupolas with simple wet arresters only or cupolas in excess of 20 tons per hour capacity. (For new cupolas the change from wet arresters to low intensity scrubbers is recommended to be at the level of three tons per hour.)

TABLE 4.11 Size of Cupola Plants

	Sample Number %	UK Number %	Proportion of UK Annual Production
Less than 2 tons/hour	2 (4)	88 (9)	0.5
2-4 tons/hour	13 (27)	390 (42)	14.3
4-7 tons/hour	23 (48)	313 (34)	27.9
7-10 tons/hour	2 (4)	50 (5)	9.4
10-20 tons/hour	5 (10)	70 (8)	23.6
20 tons or more/hour	3 (6)	17 (2)	24.3
	48 (100)	928 (100)	100.0

Source : Grit and Dust Working Party Report 1972.

In addition to these types of arrestment equipment, firms will have to comply with new recommended chimney heights (see reference (4)). The maximum discharge height for new plant will be 65 feet and most will have to be more than this due to special factors such as large amounts of steel being used in their construction or the plant being located in a valley or close to houses. In these circumstances, chimneys of 120 feet are likely to be needed and, given the substantial diseconomies of increasing heights of chimneys as indicated by Table 4.12, this is likely to substantially disadvantage the small foundry.

TABLE 4.12 Construction Costs for Chimneys

Height in feet	Capital Cost £
100	5000
120	6500
150	9000
200	14000

Source : BCIRA. These data are in 1970 prices; they may well have trebled as a result of recent rises in construction costs. The running costs of chimneys are negligible.)

Thus the Working Party implicitly recognised the special case of the small founder but, even so, as Table 4.13 shows, the capital cost of a simple wet arrester to a small founder is estimated at 50% of the cost of a new cupola itself. The very large firm will have to invest 2½ times as much in its arrestment equipment as in its cupola. Thus, because of the greater sophistication in equipment needed, the costs of emission control rise at an increasing rate with the size of the plant; furthermore the above costs do not include the considerable costs that will arise through cleaning and discharging the liquid wastes from the wet arrestment devices.

In order to assess the impact of the new controls on the industry, it is essential to know the current state of pollution abatement. The Grit and Dust Working Party estimated that in 1972 less than ten percent of all cupolas had no form of arrestment and one third of all cupolas were fitted with dry arresters which were to be outlawed.

TABLE 4.13 Relative Cupola Furnace and Gas Cleaning Costs

Cupola Melting Rate (tons per hour)	Cupola Cost (units)	Collection cost to meet Grit and Dust WP recommendation (units)	Power cost (kW/hour)
3	2	1	3
5	$3\frac{1}{2}$	3	30
8	$4\frac{1}{2}$	5	40
12	6	15	200
20	6	25	300

Source : P.J. Moseley and F.M. Shaw, The Work of the Cold-Blast Cupola Panel, National Society for Clean Air Seminar, 1975.

The results of our investigation are recorded in Table 4.14. In general the smaller firms used the dry arrester equipment and the larger firms the wet arresters, as Table 4.15 indicates. Cost was clearly an all-important factor since wet arrestment equipment would cost many times more than dry arresters; in one case the firm estimated that the cost of the wet arrester exceeded the cost of the cupola itself.

TABLE 4.14 Arrestment Equipment Used

	Number of firms
No arrestment	4
Dry arresters	7
Wet arresters	14
Other forms of cleaning	3
	32 *

* This sums to more than 29 because some firms had both types of equipment.

TABLE 4.15 Size of Firm and Arrestment Equipment Used

Size of Firm	Arrestment Equipment Used: Dry Scrubber	Wet Scrubber	Other	None
0-50 employees	6	0	0	4
51-200 employees	0	9	1	0
Over 200 employees	1	6	3	2

As Table 4.15 shows, size is an important discriminating variable and many of the smaller firms who had not the proper arrestment devices were aware that they would have to invest substantial amounts in the near future to bring their emission under control. At the same time we had the impression that, because they were often very small, and the cupola may only have been used once a week, their emissions over the whole week were probably quite satisfactory and it might be economic nonsense for them to have to fit a wet arrestment device which would only be used once or twice a week. Data collected by the Council of Iron Founders emphasises the relatively small pollution role that the small firm has. We summarise their data in Table 4.16 for production and usage of cupola by cupola capacity. It is clear from this Table that the small founder is a minor polluter in aggregate, although we would admit that he may be significant in a localised context.

TABLE 4.16 Size of Cupola and Usage

Size of Cold-Blast Cupola (tons/hr)	Annual Output (tons)	Annual Usage (hours/year)					
		Less than 250	250-500	500-1000	1000-1500	1500-2000	2000 +
Less than 2	19,000	32	35	11	2	6	2
2-3	118,000	21	26	35	26	13	11
3-4	415,000	26	57	68	26	32	49
4-7	1,040,000	21	53	63	40	84	52
7-10	351,000	-	8	12	7	10	13
10-15	716,000	2	2	7	9	12	26
15-20	163,000	2	-	2	2	2	4
20-30	498,000	-	-	-	-	4	11
30 +	410,000	-	-	-	-	-	2

Source : P.J. Moseley and F.M. Shaw, op.cit.

The small firms were clearly in an entirely different position to one firm with 300 employees we visited which had seven cold-blast cupolas, at least one of which would be running at any one point of time yet all of which were fitted only with dry arrestment. The firm's estimate for a barely adequate wet arrestment system was £60,000 (at 1975 prices). Although much of the current pollution was not, it was claimed, causing much environmental harm, it was clearly in very substantial quantities and needed something more than simple dry arrestment. We refer in Case Study 22 to an even more extreme case where a firm twice the size of the above had no arrestment at all. Thus the cost of fitting adequate arrestment would be substantial but so also would be the benefits to the community. However, to generalise on the need for legislation using this type of firm as an example but at the same time to include the small and insignificant polluters may be counter-productive.

The cost of changeover from wet arresters to high-energy scrubbers can be substantial and we have illustrated this using data from the Grit and Dust Working Party report in Chapter 2. There we observed potential economies of scale in pollution control and such economies may arise in other areas of foundry operations. First there is the potential economy through the operation of the cupola itself - capital costs will be lower per ton for larger cupolas and because the heat losses of larger cupolas may well be less than that for small ones. A second potential source of economies of scale may well be the fettling shop; less fettling may be required in a large plant operating automatic moulding equipment.

Given the existence of these potential economies of scale, it may seem surprising that small firms survive. We may explain this phenomenon by pointing out that transport costs for castings are high and there is therefore some advantage to be gained from spatially diverse plants *. Further, many plants produce specialised (even one-off) products because the small plants have the flexibility to do so and can meet the intricate designs demanded. There is also considerable ease of entry into the industry since the degree of technology required - particularly in grey iron - is low. In addition, there may be a low price elasticity of demand for castings in certain trades where the value of the castings input represents only a small fraction of the value of the final product. For example, one foundry told us that in their product line castings accounted for 80 percent of the weight of the product but only 15 percent of its value. Taken together these factors are probably sufficient to overcome the economies of scale, but if pollution control costs were to increase faster than raw material prices this would not be so.

The cost data given so far represent only one aspect of the costs firms will have to bear when they instal pollution control equipment. The loss of output that may be incurred when the plant is being installed and tested may be substantial but we have not been able to collect any hard information upon this since firms had not considered this aspect of costs. Such 'costs' are likely to be higher where there is a high degree of sophistication in the control equipment and where failure of that equipment may prevent any production at all. At the other end of the size spectrum, many very small producers exist with very old plant, some of which may never have been purpose-built as cupolas in the first place (at least one firm operates with a converted ship's boiler); many such producers told us that it would be impossible to mount simple wet arresters because the stacks lacked sufficient strength and they would therefore have to replace the whole of their cupola plant if they were to meet the emission standards. Whilst this may be desirable in some respects, we can envisage circumstances in which it would lead to wholesale closures of small plants **.

A factor that the individual firm will face when deciding about the installation of pollution control equipment is the profitability of the trade and the accessibility of finance. Naturally these are partly interdependent but they also depend upon the nature of the work undertaken in the firm and the relationship the firm has with its customers. The importance of the nature of the work undertaken arises since the iron founding industry may be more conveniently considered as two separate industries - jobbing iron founders which cast components in small batches (sometimes singly) and 'production' founders which utilise capital-intensive methods for the manufacture of long runs of standardised products. Naturally there are firms which operate somewhere within this wide output spectrum. Some firms in the industry are wholly independent but some - which we have called "captive" - form part of a larger group or organisation and these "captives" may have much greater access to capital than the independents. Table 4.17 shows how the firms we contacted can be divided into these categories; whilst nearly half the sample formed some combination of jobbing and production output, we must stress that the proportion of output accounted for in the "captive" foundries was very much lower than the independents (10 percent or less). The only captive jobbing founder had only recently been acquired by its

---

* We found one plant specialising in SG iron castings that was exporting to North America.

** It has been suggested that this is one of the main reasons for closures of small plants in the USA.

parent group and its long-term future as a jobbing founder must be in doubt. The independent production founders both mentioned that lack of capital was a constraint on their operations but, additionally, they were reluctant to tie themselves to fewer customers because they saw this increasing the risk to themselves. By comparison, the captive production founders were able to operate at higher capacity utilisations than their independent counterparts and consequently must have been able to operate more profitably.

TABLE 4.17 Production by Organisational Type

	Independent	Captive
Jobbing only	9	1
Production only	2	3
Both	6	8
	17	12

We were able to obtain estimates of capital expenditure on pollution abatement from eleven "captive" iron founders and sixteen independents. There was a considerable difference between the two groups; the independents had spent on average about £25,000 each whereas the average for the "captives" was £80,000 - more than three times that amount. Within these groups it was natural that the largest firms had spent most but, even so, one of the captives with less than 200 employees (and two others with between 200 and 500 employees) had laid out expenditure in excess of £100,000. In part, of course, the difference in the expenditure between captives and independents may be explained by their larger size but it seems inevitable that the size factor does not fully explain the differential. On the basis of the information supplied to us about their existing abatement devices and on the past cost of these devices we estimate that, out of the firms contacted in this sample, sixteen would have to spend between them approximately half a million pounds. At a time when profitability is generally low due to the economic recession, it seems inevitable that a proportion of these firms will be either unable to comply with the new air pollution requirements or unable to continue in business. This arises since, although the annualised costs may be low, the short-run cash flow impact may be considerable *.

The above paragraph suggests that it will be for reasons of the lack of finance that firms will not be able to continue, whereas of equal importance is the addition to variable costs that pollution control imposes. Most of those interviewed were unaware of the cost of pollution control equipment (22 out of the 29 did not know) and there was an almost complete lack of awareness of the power loading of collection equipment (only three knew of it). These numbers may be misleading since, given the responsibilities of those we interviewed, it may be unreasonable to have expected some of them to have known. Nevertheless, it would suggest that pollution control equipment is of small importance in total costs - for example, it may be that fuel price rises have become of much greater significance than raw materials or pollution control. Further information on this is given in Chapter IX.

---

* This is also true of dyers and finishers as will be seen in Chapter V.

As we pointed out in Chapter I, a great many factors influence the ability of an industry to pass increased costs on to consumers and the eventual size structure of that industry. One of those factors is the availability of close substitutes and in this area there is little we can be specific about since castings are used in a wide variety of applications - from gutters, to cars, to ships. Further, there are various grades of iron from ordinary grey iron through malleable and spheroidal graphite (SG) irons. Total production in the industry fell by 500 thousand tonnes from 3.718 million tonnes in 1963 to 3.281 million tonnes in 1972 *. During this period the price of castings rose 16 percent in real terms (80 percent in money terms); to a certain extent this change may be accounted for by improvements in the quality of the castings. Over this decade there were substantial changes in the usage of castings in different industries - motor vehicles and cycles increased its usage (15 percent of industry total) by 52 percent over the decade (26 percent of industry in 1972). In fact, this trade was practically the only one in which usage increased; in mechanical engineering usage fell by 32 percent or over 200 thousand tonnes and most other users exhibited similar proportionate falls. The NEDO, projecting usage to 1977, forecast few changes in market shares after 1972, with motor vehicle manufacturing remaining the largest user of castings. Here and in other sectors demand for castings was expected to be affected by factors such as technological developments of engines of new design (or even battery operated) and substitution by aluminium components. In the building trade there has been a major substitution of plumbing products by plastics and in parts of the domestic trade substitution is expected to arise from steel products. The pressure pipe industry has been affected by substitution by concrete and plastic pipe which may continue further; in addition, some fall in total volume of output in this sector has been brought about by a switch to SG iron pipes which, due to their greater strength, weigh less than their grey iron counterparts.

Imports obviously constitute a potential source of close substitutes in any market but in this sector it appears that imports represent only a minute (but rising) proportion of consumption. Exports - although very small- are also rising but the NEDO believes that export potential is limited as much by capacity constraints as anything else.

Another factor determining the ability to pass on cost rises is the structure of the industry. Naturally this is not a static phenomenon and may itself be a reflection of the industry's ability to pass on cost increases as well as a determinant of it. Between 1963 and 1972 an average of 56 foundries closed every year (net figure - there may have been slightly more gross closures) and the total number of foundries fell from 1370 at the end of 1962 to 806 at the end of 1972. As may be expected, the largest losses were in these smallest plants as the data in Tables 4.18 and 4.19 demonstrate.

It was suggested to us that these changes in the structure of the industry may slow down now that excess capacity of the jobbing founders has been reduced and the limits of substitution have been reached for the capital-intensive foundries. There is slight evidence available of such a polarisation occurring but not of a slowing down of the pace of change, so we must reserve judgment on this for the present.

---

* These and subsequent data on the industry are drawn from the Industrial Review to 1977, published by NEDO (5).

TABLE 4.18 Number of Iron Foundries at Year End

Employees	1963	1967	1971	Losses 1963-71
1 - 50	822	615	497	325
51 - 200	351	307	248	103
201 - 500	96	88	73	23
501 - 1000	28	17	15	13
1001 +	9	10	10	-1
Total	1306	1037	843	463

Source : NEDO

TABLE 4.19 Proportion of Output in Different Foundries

	Number of Foundries		Output	
	1963	1971	1963	1971
Annual production tonnes	%	%	%	%
Less than 1200	69	59	10	6
1201 - 5000	21	25	19	16
5001 +	10	16	71	78
	100	100	100	100

Source : NEDO

One further factor suggested to us was that there has been a tendency for consumers to specify castings of even higher quality and therefore those producers able to supply high quality castings will be more able to survive in the future than those not so capable. Indeed, one firm mentioned that they felt that had it not been possible to change over to the production of SG iron, it would have gone out of business. The new equipment installed to enable this to be done (an electric induction furnace) also brought an additional benefit by substantially reducing the air pollution output of the factory (apart from times when oily scrap was being processed). Such benefits are not only limited to the larger producers; one firm employing no more than ten people had changed over from an oil-fired furnace to an electric furnace with a resulting reduction in pollution output. The change had been stimulated by a lack of spare parts for the oil-fired furnace; in no way, therefore, was pollution a primary consideration in these two instances but a secondary benefit derived by the firm and community from technological change and economic factors.

In addition to external air pollution caused by the melting process, we must also mention the other sources of external and internal air pollution by dusts and fumes. Fumes can arise in two ways; electric furnaces generate hydrocarbon fumes if oily scrap is melted and the firm that mentioned this told us that it had spent much effort in collaboration with the authorities investigating this problem. It seems that little can be done without the installation of expensive degreasing plant like that operated by firms in the non-ferrous trades. Secondly fumes arise when molten metal is poured into the casting boxes; in the past this has not been a great problem but the technological changes that have given rise to the use of resin-bonded sands has brought with it a substantial fume problem in the casting shop. The offensive odours cannot be vented to the atmosphere if the plant is located near domestic dwellings and one firm we spoke to is actively considering the installation of a supplementary furnace to burn the gases given off.

Dusts can cause problems in various parts of a foundry but are probably most acute in the fettling and sand reclamation plants. Almost all foundries carry out some degree of fettling and there are specific legal requirements which have been laid down to cope with the particularly nasty conditions that may be encountered in fettling shops. Simple face masks are not really adequate and a more sophisticated approach must be adopted such as the use of specially designed fettling booths with their own integrated air extraction system. Such booths are really only suitable for small castings and other methods may have to be designed to deal with larger castings (see, for example, (6)). Internal dust control is not cheap, as can be seen from Table 4.20.

TABLE 4.20 Cost of Dust Control Equipment in Foundries

Size (number of employees)	Number of firms	Average cost of equipment in the group (£000)
Less than 50	6	11
51 - 200	9	26
201 - 500	3	142
501 - 1000	1	750
1001 or more	2	225
Total	21 *	92 (overall weighted average)

* Not all firms were able to give us information on these costs.

An average figure of approximately £90,000 was indicated by the foundries we visited which were able to give us the data and, of course, those firms will have higher operating costs through increased power consumption. There will probably be some tangible benefits to off-set these costs (in part at least); a dust collection system may enable more sand to be recycled thus saving on this and there may well be substantial savings due to lower absenteeism through reduced accidents and disease. Fettling is a very dirty trade in which unskilled workers earn low pay with a consequent high labour turnover; cleaning up the conditions in the fettling shop could well lead to lower labour turnover thereby saving on the costs of labour procurement and training.

Iron founding is a traditional industry and has been located in parts of the East Midlands for a long time. This traditional and historical aspect we found to be quite important from both the internal and external environmental viewpoints. Inside the foundry considerable health and safety steps had been taken but conditions were still unpleasant; generally we found that old-established firms and the older workers in all firms were more likely to accept the current internal conditions whereas newer firms and younger workers were the ones who typically found some aspects of the trade still difficult to accept. We believe this perceptive bias was also true for the way in which local authorities viewed foundries. Thus in Case Study 19, the firm felt it had come under undue pressure because it was not in a traditional founding area, whereas in many other instances we found evidence to suggest that the traditional role and aspect of the foundry was crucial in a local authority's leniency towards pollution.

The foundry industry presents a fascinating microcosm of how industrial market structure, technical progress and the processing of information by those within and outside firms can all play an interactive role in the history and future of the industry. In the context of our research there are clear problems for the small firm but the medium sized firm that is not integrated into any larger group may find that pollution control presents an even more serious threat to its survival.

Dyers and Finishers

Textile dyeing and finishing is usually regarded only as a major source of liquid effluent, yet an analysis of replies in our stratified sample of dyers and finishers showed that, in some instances (albeit a minority), there may be substantial air pollution problems to be overcome. In three cases the costs of air pollution control exceeded that for control of water pollution, although it must be admitted that these cases were exceptional due to a unique combination of factors such as location, process, raw materials and the like. Our evidence would indicate that, unlike most, in these cases the output of pollutants per unit of output seemed positively related to economic growth, progress and innovation. On one hand we have the use of synthetics and solvents and on the other the use of more sophisticated monitoring equipment by local authorities extending our knowledge of the nature of pollutants.

The principal airborne pollutants could be classified as follows; stenter fumes, phenolic fumes, chlorine fumes, sulphur fumes, water vapour, sulphur dioxide, carbon dioxide (and all normal products of combustion) and airborne textile fibres. The whole methodological problem of allocating pollution control costs jointly between productive and unproductive inputs is quite well illustrated in this particular area since many of the fumes were discharged to the outside environment in order to improve internal working conditions. Thus a chimney may be needed to provide downdraught for combustion processes, to make the shopfloor a more pleasant place to work in and to provide dispersal of pollutants around the factory (i.e. pure pollution control). Problems of joint cost allocation of this nature were difficult to overcome and, in many instances, we have to make crude estimates based on qualitative information.

Forty firms in the trade were contacted and thirty-two were able to give an interview. We are satisfied with the coverage of our sample although we believe that one of the firms not replying to our enquiry was a major user of pollution control equipment. Generally, with this single exception, our sample was able to cover the widest variety of firms defined by size, organisational structure (i.e. whether or not part of a larger group either on or off site), product type (synthetic or natural fibres, hose or yarn), process type and geographic location. Some of these parameters are illustrated in Tables 4.21, 4.22 and 4.23.

Several points emerge from these tables but two particular ones should be made explicit. First, there is a preponderance of dyers and finishers in heavy urban areas, and this could

TABLE 4.21 Geographical Distribution of Dyers and Finishers

Location	Number of Firms
Nottinghamshire	16
Leicestershire	14
Derbyshire	2
Lincolnshire and Northamptonshire	0
	32

TABLE 4.22 Nature of Location of Dyers and Finishers

Location	Number of Firms
Rural	0
Urban	9
Industrial estate	7
Heavy urban	16
	32

(Note : these classifications are subjective but correspond to those given earlier in Table 4.8.)

TABLE 4.23 Size and Organisational Structure of Dyers and Finishers

Number of employees	0-50	51-200	201-500	501-1000
Organisational type :				
Independent	3	11	1	1
Captive and/or integrated	1	10	2	3
	4	21	3	4

place space constraints costs * on the firm if pollution control necessitates any major change in process. From the point of view of air pollution, the location of the firm determines the harmfulness of the wastes since it may facilitate or exacerbate the dilution of emissions before they impinge upon areas of human habitation. Secondly it is important to note that this is an industry where two distinct types of firm operate; there are those which are merely adjuncts of a larger organisation and there are those which take on work on a commission basis from other unrelated textile manufacturers. (To a certain extent there is a quasi-vertical integration since a dyer may not be owned, but may be virtually controlled by a big customer such as Marks & Spencer, or a big supplier such as Mansfield Hosiery Mills.) Notwithstanding this, we can say *a priori* that firms in the first category might be able to withstand the costs of pollution control less easily than those in the latter

* That is, the firm may incur higher costs or charges if it does not have sufficient space to be able to construct (for example) its own pretreatment plant. Of course it may be that, even if space were available, planning permission may not be granted.

category which are typically larger and for whom there are always possibilities of short-run cross-subsidisation to enable them to overcome the short run cash flow problems generated by pollution control. To our knowledge only two firms we came across were about to or had just closed down and both were of the small, non-integrated type. In a large number of other instances our respondents volunteered the view that the distinction between independent or group ownership was particularly important.

Hot water is an essential prerequisite in practically all dyeing processes and all firms emitted air pollutants of the usual type caused by the burning of fossil fuel in their steam-raising plant. As Table 2.24 shows, the use of coal/coke products was minimal so the majority of airborne pollutants in the industry were essentially of the less visible type.

TABLE 4.24 Method of Steam Raising used by Dyers and Finishers

	Number of Firms
Coke	3
Gas	7
Oil	13
Gas/oil	9
	32

Oil usage varied widely from 2000 gallons to 20,000 gallons per week but it made little impact on the surrounding environment, although in several cases our respondents stated that local authority pressure had forced them to use low sulphur oils which cost up to 25% more than regular grades. In only four cases had a switch from coal or coke been made on anything but grounds of efficiency and to a large extent our respondents were unaware whether their plant was located in a smoke control zone. A much larger number (20 firms) had had to change their processes and emissions in some way because of local authority action under the Public Health Acts. This was often because noxious fumes were from time to time being emitted from parts of the dyeing process. In 17 of these cases the major problem was identified by the firm as stenter fumes * and in all cases these were some organic emissions from dyeing machines. The worst emissions arose from waft knit synthetic fibres. Synthetics (and in particular polyester) require high heat setting temperatures and at the setting stage small quantities of organic products such as spinning and knitting oils, dye carriers and resins remain in the material; some part of these is volatised in the stenter and exhausted to the atmosphere. The quantity of emissions depends on the quantity of residual organics and, in the case of knitting and spinning oils, this in turn depends on the amounts of oil originally added and the efficiency of scouring or other removal methods. As far as dye carriers are concerned, emissions depend on the quantities used so dark shades are worse than light shades, although recent innovations in high temperature dyeing have resulted in less carrier being used; although this is a more costly dyeing technique than the traditional ones, its use is more a reflection of quality demands from the market place than the need for pollution controls. There were indications that stenter fumes may be both positively and negatively related to the quality of the product, so we can make no unambiguous statements whether pollution is the price that must be paid for better quality products.

---

* Stentering is the process by which, with the application of heat, fabric is flattened after dyeing.

Two major problems of perception arise with these fumes; they are usually visible (depending on the carrier or oil used) and they have an unpleasant smell. Thus, when a dyeing firm is situated close to dense housing, it is likely there will be complaints even though the physical impact on the environment may be small. Severe and recognisable stenter problems were rare and, although most firms employing the waft knitting process had some problems, only four were substantial. Products which led to the emission of large quantities of fume were frequently only about 10% of the firm's output but in those few instances where the problems were severe, very difficult control problems were encountered. Table 4.25 details how firms in our sample dealt with the fumes.

TABLE 4.25 Method of Dealing with Stenter Fumes in Firms where they Arose

Wet scrubbing	1
Catalytic oxidisation	2
High dispersal	11
Dust extraction (internal)	17
Other *	4
Total number of firms	17 **

* Transferred operations to another factory or changed carriers or changed materials/dyes.

** Total adds to more than 17 because some firms used more than one method.

The removal of these fumes from the working environment is most important; this has been recognised by the authorities and the Factory Inspectorate is, we are told, beginning to take a harder line here. Of course, dust and fume extraction merely transfer the problem to the external environment, so high dispersal of the fumes through chimneys of 60 feet and above is necessary. If this is still unsatisfactory, the fumes can be cleaned by wet scrubbing but this causes problems because much of the fume is insoluble and also because large amounts of knitting oil contaminants tend to condense out in the exhaust trunking before reaching the scrubbing device. The other approach to the problem is potentially difficult and expensive; the fume may be incinerated at temperatures of 750°C. This clearly involves substantial fuel costs and is technically hazardous. Alternatively, it is possible to use what is called catalytic oxidisation; this has considerable fuel and other cost savings over the first method but the technology is still immature and maintenance costs are high. Finally, changes in processes and inputs can be made; this is so far an unexplored area and in the four cases of this type that we noted, there was some ambiguity as to whether process changes had occurred because of pollution or because of some exogenous product/market need.

Stenter emissions are clearly widespread in the industry, yet when asked open-ended questions about fumes, only four firms volunteered a reply that contained some reference to stenter fumes. This may merely be a reflection of this type of pollution in the sense that it may not constitute a 'bad' unless wind direction and proximity of domestic dwellings are combined in specific ways. These circumstances may, fortunately perhaps, be very rare and a great deal of research may be needed to isolate the problem. Our conversation with the Leicester Local Authority suggested that it was only after a large number of complaints had been received (some about distorted vegetation) that one firm's stenter fumes were found to be causing a nuisance and even then it took two years to isolate as the cause one dye which happened to have herbicidal properties yet was used only infrequently. The action of this local authority helped to identify and reduce the problem and such was the paucity of information that their efforts eventually became internationally known. In other areas local authorities have so far taken little action; indeed, only Nottingham and Leicester seem to have any real perception of this problem. One firm commented that, on some things as complex as this, the local authority made only token noises when confronted with complaints. Our knowledge of the eco-effects of the fumes are still sparse; a few well-documented case studies suggest it can lead in some circumstances to distorted vegetation but it appears that the main harm emanates from the odour. It is because so little is known about these odours that the technology of control is still in its infancy. Another firm remarked to us "exhaust fans have been installed above the stenters because of complaints from operatives; the problem is we don't know how harmful the fumes are".

Six of the firms emitting stenter fumes did nothing more than ventilate the fume out of the factory although one was situated close to housing. Where some dispersal method was used this mainly took the form of a high chimney although it was not clear that the disposal of stenter fume was the primary purpose of the chimney. Few firms recognised stenter fumes as a major source of nuisance.

Where we were able to attribute costs to the disposal of stenter fumes we found that the capital sums amounted to no more than a few thousand pounds, but there were instances where, for example, the cost of the chimney could be wholly attributable to the stenter process and in one such case the firm said it had spent over £10,000 on its high chimney. As we have already noted, some firms had changed or phased out processes to avoid the stenter fume problem altogether; this is not without its costs. One firm (later to go bankrupt) claimed that it discontinued making an elastic fabric which previously accounted for one third of its turnover and two firms adopted the relatively new and untried process of catalytic oxidisation. The cost for each stenter machine was about £11,000 for the oxidisation plant and £7000 for the necessary adjuncts. In one firm the total capital cost of disposing of stenter fumes was £54,000 yet its turnover was £220,000 per year and its profit on sales was about 8%; it is easy to see how pollution control costs can have serious financial implications for firms. This firm was faced with a substantial investment in capital equipment which would have no positive output effect because the planning authorities had objected to a proposed chimney and the Factory Inspectorate objected to other alternative disposal methods. Another firm we interviewed was potentially faced by a similar position; over the past two years it frequently had to alter its processes due to the regular complaints of a small number of people on a nearby housing estate. In the event, we interviewed the firm's manager just a few days after it had gone bankrupt as a result of the low margins prevalent in the industry.

One of the firms which had installed and tried the equipment had found it needed a lot of maintenance, partly because their product had a brushed fluffy finish and the fluff had been sucked into the oxidiser and caught fire. Eventually, after agreement with the local authority, it was decided that, in addition, changes in processes and product should be made. Thus in these cases pollution control costs are virtually double the estimates we would make if we were to take into account water pollution charges alone and they also illustrate how pollution control is dogged by problems of uncertainty, lack of information on the pollutant and its reduction, and variability of enforcement. The latter depends on a variety of factors from number and type of complaints, to the ability of the relevant local authority to pinpoint and specify the source of pollution.

So far the costs we have considered are essentially tangible costs and there are, of course, a number of costs which are just as important to the firm but which may not be readily apparent. Plant stoppage, management time to deal with complaints and changes in processes and materials are three areas where there may be very real costs of pollution control. Even when a pollution control device is fitted, it may cause a problem in another area (we found this was particularly true of some air pollution control devices which can cause noise problems).

About half the respondents in this industry felt that there would be tightening up of their airborne emissions in the future. This was because they considered fume problems were becoming better understood by local authorities and the action that could be taken on them was now clearer. About two-fifths of the sample said they were considering changes in processes or utilising some form of scrubbing device in anticipation of changes in regulations.

Control of the internal environment in the factory may also necessitate substantial expenditure; in all cases some form of ducting or face masks were used, but in only 19 cases was the expenditure incurred substantial enough for our respondents to take note of it. In most cases where respondents knew of the costs they were capital expenditure on ducts and fans which could cost upwards of £1000. Table 4.26 details the ranges of expenditure reported to us.

TABLE 4.26 Internal Fume/Dust Arrestment in Textile Dyers and Finishers

Capital cost of equipment (£)	Number of firms
Less than 1000	5
1000 - 3000	3
3000 - 6000	5
6000 - 9000	1
9000 +	4

In the majority of cases the equipment was for one of four main purposes; to extract stenter fumes, solvent fumes, dye particulates and loose fibres. The main dangers were to a comfortable working environment rather than any direct health risk, although in the case of dye particulates and solvent fumes there were possible dangers of cancer-related diseases. Loose fibres were a problem but there was nothing like enough to cause the byssonnossis diseases of the Lancashire cotton industry. Most of the costs were for improving working conditions and would (hopefully) have some recognisable impact in terms of better output; they cannot therefore be considered as pure pollution control in the way the other equipment discussed in this chapter can. In a similar vein, most companies spent considerable sums of money on equipment to reduce condensation in the factory but, in essence, this was to protect the structure of the buildings and machines and was in no way a pollution control device. The internal pollution problems were not of a really serious nature and in only two cases were we told of equipment being installed on the instructions of the Factory Inspectorate; generally it appeared to be a management or joint management/trade union decision to improve working conditions.

The Role of Local Authorities

Local authorities play a dual role in the control of air pollution which is characterised by its positive and negative aspects. The positive role is in terms of the role of the local authority in establishing smoke control zones, in the vetting of planning applications with regard to their influence on the environment and in a continual dialogue with industry on its control measures. The negative aspect, on the other hand, lies in the enforcement of controls like the 1956 and 1968 Clean Air Acts, dealing with complaints from the public about air pollution and where, and when, necessary ensuring the prosecution of offenders. Most local authorities also undertake the monitoring of air pollution by measuring, for example, smoke and sulphur dioxide concentrations in various parts of the district (although few monitor anything else). Thus it can be seen that the local authorities have a wide range of functions within the area of pollution control although probably the single most important positive function they fulfil is in the setting up of smoke control zones. Whilst these largely affect domestic households only (they often deliberately avoid industrial areas) and thus their impact on industrialists is slight, we justify the inclusion of reference to the local authorities in this report by the fact that pollution control costs are borne by society as a whole, as well as industrialists in particular, and by the fact that they operate many of the pollution controls.

We contacted twenty-nine out of the thirty-three authorities in the area; those omitted were largely residential in nature and thus did not come within the scope of our enquiry. The degree of cooperation we obtained was very good - in none were we refused an interview. We have attempted to classify the areas according to their degree of industrialisation and Table 4.27 shows our subjective assessment here; it can be seen that nearly one half of the councils interviewed were highly industrialised. It will be seen later that there were substantial differences in the resources available in the different types of area.

TABLE 4.27 Nature of Industrialisation in Councils Visited

	Number
Wholly or largely industrial	7
Essentially industrial - some agriculture	6
Essentially agriculture - some industry	8
Industry and agriculture mixed	4
Mainly residential - some industry	4
	29

NB - Classifications are entirely subjective.

Pollution control in a local authority is more often than not within the hands of an "Environmental Health Department" which was often under the direct control of the local Medical Officer of Health until local government reorganisation; since then most have become separate departments. The result of this was that the controlling committee was responsible for a wider range of activities than merely pollution control - for example, food, hygiene, rodent control and slum clearance were often amongst the functions within the control of these committees. Furthermore, since the local authorities are in some ways connected with all aspects of pollution (solid waste disposal, water pollution in the recent past and noise - in addition to air pollution) it is almost inevitable that for some while the local authorities have been hampered by a lack of manpower with specialist skills in the area of air pollution yet, ironically, it has been here that most has been achieved. Table 4.28 shows the extent of smoke control zones designated until the end of 1974 and the premises they cover. It can be seen that the largest number of "black areas" exists in the GLC and most progress has been achieved there. The East Midlands comes out badly; less than one third of all potential areas had been designated by the end of 1974 although these areas did cover two-fifths of the total number of premises. Overall nearly two thirds of all premises are covered and this is an impressive achievement which has been instrumental in bringing down smoke and sulphur dioxide concentrations quite dramatically.

TABLE 4.28 Progress of Smoke Control Zones to 31.12.1974

Region	Total Number of Black Areas	Proportion of Area Covered %	Proportion of Premises Covered %
North	45,690	36.4	32.6
Yorkshire and Humberside	205,304	54.5	59.0
East Midlands	73,703	27.5	43.6
West Midlands	92,512	37.1	40.5
North West	211,265	52.6	51.8
South West	7,505	28.5	19.3
G L C	260,582	79.7	85.8
Whole Country	896,561	50.6	60.4

Source : Clean Air, March, 1975.

The local authorities are not, however, entirely free agents to do as they wish. For example, in 1954 the Beaver Committee, concerned at the slow progress of smoke control zones, established the so-called (and now defunct) "Black" and "White" areas and it was in the former that highest priority was to be given to smoke control. Similarly in the case of pollution control through planning *; consultations between planning departments and environmental health departments over the pollution implications of planning implications were a de facto reality in many authorities but the DoE circular 10/173 now requires this consultation over the matter of noise pollution. There are many other instances in which the DoE can be seen to be shaping, if not directing, the pollution control work of the councils. This control is not necessarily in a positive direction; with regard to the recent financial stringencies, the government has suggested that plans for smoke control zones

* For a fuller discussion of this topic, see reference (7).

could be pruned and, in another area, noise pollution controls would have to be provided out of existing resources. The control by central government is two fold - the direct control operated by the issuing of directives, guidelines and standards, and control through finance (only 40% of the cost of smoke control zones is provided from local sources, 30% being provided by central government and the rest coming from the householder).

As has been intimated above, it is the responsibility of central government to set down standards and issue guidelines; local authorities do not have the technical expertise to do this for themselves. Centrally imposed standards also have the merit of tending to induce a degree of uniformity over the whole country. Local authorities do, however, have fairly close links with other advisory centres such as the government laboratories, although one authority had made an approach to the Warren Springs Laboratory for an appraisal of resources and policies only to be told that it was an "internal matter".

Apart from the setting up and maintenance of smoke control zones, one of the principal functions of the local authority pollution control section lies in dealing with complaints and, in some cases, the prosecution of offenders. It is hard to present any meaningful statistics here since the concept of a single "complaint" is hard to define. One individual complaining every day for a month may constitute either one or many "complaints" in the records of an individual authority. Thus the collection of comparable data is probably impossible. Some authorities in the area - such as Beeston - had, in fact, attempted to study the awareness of members of the general public of air pollution. These studies tend to confirm others conducted in the US * that, although many people claim awareness of pollution and pollution control, it is only a relative few who do anything about it. Our contacts in local authorities did, however, tell us that although there was little hard evidence on this it seems that people who had recently moved into an area were more likely to complain than those who had lived in the locality for a long time. Further, most complaints arise only in those areas where authorities are empowered to act. Many authorities see education of residents about pollution control laws as one of their duties. One thing stood out from our enquiries; there are far more complaints about air pollution than there are prosecutions. The authorities seem to rely on education and persuasion and resort to prosecution only rarely in extreme cases (due, in part, to the amount of time prosecutions take up). For example, in the City of Leicester there have only been, on average, two court actions per year concerning air pollution in the last six years. In one case a statutory notice was served on a foundry which had delayed installing wet arresters and, three days after being served, the foundry agreed to have them installed. In another case a dyeing company was served notice under the Recurring Nuisances Act to :

(a) properly insulate or line the chimney to a full height and insulate the flue system along its full extent to minimise heat loss ;
(b) undertake such other steps as may be found necessary to abate the nuisance including substitution of a fuel oil with a sulphur content lower than 1%; and
(c) maintain and operate the plant in an efficient manner.

It is clear that the number of complaints to or prosecutions by local authorities is no reliable guide to the effectiveness of air pollution control measures. Local authorities tend to regard prosecution as a measure of last resort and prefer persuasion to coercion. Further,

* See (8).

there is some evidence that, paradoxically, the number of complaints about air pollution has increased with the introduction of smoke control zones *. Evidently the public, forced to change their traditional heating methods, have become more aware of smoke emissions whether they occur inside or outside the smoke control zones. Thus the number of complaints is a measure of the success of the smoke control policy; although we initially tried to collect information along these lines, it soon became apparent that its usefulness was so limited as to make it not worthwhile. In addition, analysis of complaints to Nottingham Corporation showed that in 1968 only 1.7% of all complaints concerned smoke, grit, dust and fume and only 2.5% concerned offensive odours. In 1972 these figures were 0.6% and 1.8% respectively. This suggests that industry's air pollutants cause a small and diminishing number of complaints to local authorities.

A further aspect of the complaints system that should be mentioned is that until recently at least there has been an almost total reliance by local authorities on complaints from the public for their information on likely sources of pollution. In very few instances did we find local authorities taking the initiative in air pollution control - partly as a consequence of inadequate staff and partly as a consequence of their limited powers. In many ways this is unsatisfactory; relying on the public who may not have sufficient technical knowledge to be able to detect a potential health hazard is undesirable to say the least.

The very diversity of the nature of the environmental health departments' work makes an assessment of the amount of resources devoted to pollution control difficult; this is now exacerbated by the need for consultations with planning departments. There is an increasing tendency for authorities to employ specialist smoke control officers but it remains true that the total amount of resources devoted to pollution control by local authorities is largely determined by the chance interaction of the power and interests of members of the local council and of the head of the Environmental Health Department. Further the enthusiasm with which smoke control zones go forward is partly a function of local politics; it is also true that the more northerly coal-mining areas of the region find it hard to press on with smoke control zones at the rate pursued by towns like Leicester, say. Naturally some localities - such as small towns in the country - do not need to allocate large resources to air pollution control. This is evidenced by Table 4.29 which shows the proportion of councils of each type using air pollution monitoring equipment and employing specialist air pollution staff.

TABLE 4.29 Incidence of Monitoring Equipment and Specialist Staff

	Monitoring Equipment %	Specialist Staff %
Wholly or largely industrial	84	84
Essentially industrial - some agriculture	100	50
Essentially agriculture - some industry	75	25
Industry and agriculture mixed	100	100
Mainly residential - some industry	75	-
All authorities	86	50

* This negative association between complaints and air pollution was dramatically demonstrated in (9). The graph shows a clear inverse relationship between complaints and solid deposit rates over the period 1945-1970.

Further, American studies have shown little relationship between the severity of pollution problems in particular localities and the amount spent by local authorities for control purposes; we are sure the same is true in this country. When we asked the authorities about the sums they spent on pollution control, they found it very difficult to reply and there was naturally a very wide dispersion in the value of the figures we received. For example, Derby Corporation were able to give a fairly detailed breakdown of the staff employed on air pollution control; one pollution officer was devoted full-time to this matter and about one tenth of the time of two other officers was so engaged. Including clerical assistance, we estimate that the salary bill in 1974 for pollution control work was £7,500. Of course not all their time was devoted to industrial air pollution control but it seems likely that such work accounted for just over half of the time spent on air pollution control; we thus estimate that the salary bill for industrial air pollution control can have been no more than £5,000 in 1974. Of course, to this should be added the - presently very small - element of the cost of work carried out in planning departments in consulting over pollution matters. In addition to the labour input of pollution control - apart from the costs of servicing that labour by way of accommodation, heating, telephones and so on - must be added the cost of air pollution monitoring equipment and the cost of establishing smoke control zones. The capital cost of monitoring equipment was very small; in no council did it exceed £1000. On the other hand, expenditure on smoke control zones had been substantial in some authorities; for example, examination of the accounts of Chesterfield Borough shows that up to 1973 their capital expenditure on six smoke control zones had been £50,000, to which should be added of course the central government's contribution which is almost the same amount again. Whilst not inconsiderable, these sums must be counted as very small in comparison with the other activities that local authorities undertake. Again, in Chesterfield Corporation, expenditure in connection with the Clean Air Acts in 1972/3 was £4000, compared with a total budget of the Medical Officer's Department of £56,000 which compares with an expenditure on Refuse Collection and Disposal of nearly £150,000. Also the capital expenditure on smoke control may be compared with expenditure on, say, housing which runs into millions. A comprehensive survey of all the authorities in the area has not been possible; authorities publish their accounts in varying degrees of detail and local government reorganisation has not helped. We believe that authorities should publish their accounts in such a way that pollution control expenditure (as opposed to the more general heading "Environmental Health") can be more readily isolated.

We asked the local authorities what they consider the most pressing problems in pollution control. Many told us that the lack of skilled specialist staff had held them back in the past; pollution control is now such a complex matter, requiring training in many aspects of science, that it is no longer possible to learn the necessary skills "on the job" and more formal requirements are now sought. Councils felt that such skilled people were now arriving on the job market but concern was expressed at the salaries that were having to be paid as a result of the shortage. Authorities also expressed concern at the stop-go nature of policies adopted by central government; they were concerned that generally pollution and pollution control seemed to come low in the order of priorities and always seemed first to be cut in a recession, although the Control of Pollution Act and the new planning provisions entailed a greatly increased work load. This resulted in a lack of cohesion and cooperation between local authorities, central government and other government departments that could have been avoided by greater constancy of application.

Many authorities told us that their position over industrial pollution was not unequivocal; whilst also being the guardians of the residents of the locality they also recognised the need to provide the people with jobs. Indeed, it was intimated that in many areas of higher unemployment the provision of jobs would take precedence over pollution abatement. This interface between people, planners and producers has been met in Coventry by the setting up in 1971 of a Pollution Prevention Panel consisting of representatives of industry, local authority, Factory Inspectorate, a conservation society, Warwick University and Lanchester Polytechnic (10). Such a Panel seems to us an ideal way of catering for the needs of the local community whilst ensuring the continuance of local employment.

The traditional relationship between the local authorities and industry is changing in another fundamental way; the air pollution provisions in the 1974 Control of Pollution Act - soon to be fully implemented - empower the authorities to measure and record emissions into the atmosphere and also serve notices requiring premises, including scheduled processes, to modify their emissions.

A further problem the local authorities had encountered was the aftermath of the local government reorganisation in 1974. This left new areas with the task of working out a coherent pollution abatement strategy for the whole area and inevitably led to some unevenness of the type and standard of abatement services provided in some places. The reorganisation did, however, help to work in the opposite direction; the new, larger authorities were able to bring together all pollution control skills under one roof and thereby create a new, more highly trained and specialised abatement force.

To conclude, whilst we feel that the local authorities play a large part in pollution control, their main impact and main efforts are in the area of domestic smoke control. The resources they devote to industrial air pollution are not large in comparison with their other activities and their principal line of approach seems to be one of persuasion rather than coercion. The few prosecutions for contravention of air pollution laws should not be seen as a sign of failure of the policies.

Conclusions

Pollution of the atmosphere is, perhaps, the area of pollution of which the layman is most aware and we have therefore given the subject wide coverage. However, our conclusion is that, except for comparatively small localised areas, domestic coal burning is responsible for the largest proportion of pollution and the impact of small firms on the environment is generally quite low. There are a few trades for which this is not true - particularly metal melting - but these trades account only for a small proportion of total industrial output.

We find that in the main firms have not been adversely affected by smoke control zones - many installed non-smoke emitting plant in advance of impositions of controls for technical or other reasons. The metal melting trades have been - and are being - required to provide more extensive pollution abatement equipment and we find evidence that the nature of the controls required is not neutral between firms of different sizes and we anticipate that the controls to be introduced will have an impact on the number of firms in the industry and its structure.

Our enquiries have shown that in many cases the sums that have been spent by firms on external air pollution control are quite small although this statement must be strictly qualified. First, although some firms actually achieve a profit out of pollution control, there was a marked tendency for firms to underestimate their pollution control expenditures. Second, in those firms where substantial sums had been spent on pollution control, the short-run cash flow effects of the expenditures can be considerable even if the annualised costs are small; these short-run costs are exacerbated further when finance for output-expanding projects is in short supply. Thus pollution control expenditures may lead to a slowing down in the rate of growth of output. Third, the annualised costs of pollution control equipment should also include the rates paid by firms on such equipment; all firms in our survey ignored this and we would like to see some way of reducing this penalty on firms which results in the community benefiting twice over from pollution control.

Further we find that, if anything, firms have spent larger sums on the control of the internal working environment than on control of the external environment. Whilst we would not consider this as part of pollution control proper, it does put into perspective the sums that are being spent on pollution control.

Finally we find that the resources devoted by local authorities to industrial control are small but growing. Attention has in the past been focussed on domestic air pollution through the creation of smoke control zones but authorities are gradually increasing and developing their technical expertise in the field of industrial air pollution where the hazards may be greater and the technical problems greater still.

References

1. D.J. Harris and J.F. Garner, Environmental Pollution Control, Allen and Unwin, 1974.

2. R.M.E. Diamant, The Prevention of Pollution, Pitman, 1973.

3. The Solicitors' Journal, 13th June, 1975.

4. Ministry of Housing and Local Government, Chimney Heights; Second Edition of the 1956 Clear Air Act Memorandum, HMSO, London, 1967.

5. National Economic Development Office, Industrial Review to 1977 : Iron Founding, London, 1974.

6. BCIRA Broadsheet 29, BCIRA, Alvechurch.

7. N. Lee and C. Wood, Planning and Pollution, Journal of Royal Town Planning Institute, Vol. 58, 1972, pp.153/8.

8. D.E. Fromm, F. Probald and G. Wall, An International Comparison of Response to Air Pollution, Journal of Environmental Management, 1973, pp.363/366.

9. G. Wall, Air Pollution in Sheffield, International Journal of Environmental Studies, Vol. 5, 1974.

10. R. Stiles, Prevention Rather than Cure, Trade and Industry, 7th March, 1975, HMSO.

Chapter V

WATER POLLUTION

The Present Legislation

The discharge of trade effluent into public sewers has until recently been largely under the control of the local authorities; since these were reorganised in 1974 control is now exercised by the relevant water authorities which are responsible for both supply of water and disposal of effluent. Only those firms which discharged trade effluent into a public sewer before 3rd March 1937 were not subject to this control and then only if the trade effluent was unaltered in nature, composition and quantity; this exemption was removed by the Water Act of 1974 giving exclusive control to the water authorities. New discharges after that date are allowed only if conditions laid down by the local authority (now area water authorities) are met; such conditions may be :

(a) degree of acidity or alkalinity,
(b) quantity of effluent,
(c) water temperature,
(d) time of discharge, and/or
(e) the removal of substances prejudicial to the sewage treatment process (such as arsenic and cyanide salts which may poison the bacteria used for the breakdown of sewage).

Further the Public Health Act of 1936 * makes it a criminal act to put any of the following materials into a sewer :

(a) matter likely to harm the sewer, restrict its flow or prejudice the treatment process,
(b) any hot chemicals or water (in excess of 110°F) or chemicals which, in combination with the contents of a sewer, may produce a health hazard,
(c) petroleum spirit or calcium carbide.

Further regulations govern direct discharges into rivers or the sea, but since most firms we interviewed did not dispose of their effluent in this way we will not elaborate these details. The 1974 Control of Pollution Act tightens some of these controls, allowing for a register of consents and analyses ** to be open to public examination unless exemption is granted in order to protect a trade secret ***. Further, relatively minor, changes are described in (3), including the requirement to meter and monitor discharges.

---

* Further details may be found in (1) and (2)

** Analyses were previously not available to the public.

*** The Severn Trent water authority has already published a list of their consents and analyses together with the names of those organisations which refused to give their permission for this to be done; this quite successfully anticipated the legislation and provides a base by which any future improvements in water quality may be judged (see 4).

Results of our Survey

Water pollution arises to some degree from almost every industrial process and, as with air pollution, it is necessary to estimate emissions in terms of quality as well as quantity of outputs. Using both scales there are enormous differences in the outputs between and within industries; even in an industry where all firms use the same process, substantial disparities still emerge. The harm that the discharges do to the environment (and arguably the charges or controls that they should be subject to) is unfortunately not directly calculable from output data alone, no matter what scale is used. There are many instances of firms with substantial discharges but since they are relatively distant from other sources of pollution their marginal effect is small. Some times one firm's discharge may help balance out and neutralise the discharges of others; thus frequently in dyeing and finishing alkali and acid discharges effectively neutralise each other. Some substances may have a positive benefit to the aquatic environment in small quantities; for example, it is conceivable that the discharge of fluorides may obviate the need for authorities to add that substance for the purposes of dental care. Neither can we always be certain that bio-degradable substances are preferable to non-biodegradable ones. In some instances very large discharges of organic matter will lead to massive growths of aquatic vegetation which may ultimately use up all the oxygen in the water course and actually prevent other animal and fish life. One is thus operating along a pollution continuum where the trade-off with the wellbeing of society may have a changing and non-unique sign. Thus the aggregate data which are quoted below cannot necessarily be used as evidence for comparison with other regions, nor can the firm and industry data that are used later be taken as unambiguous magnitudes in all circumstances.

In our survey liquid pollution occurred as acids, alkalis, oil, colourants, suspended solids, toxic and thermal forms. The raw effluents (although in some cases firms operate pre-treatment plants) were discharged into rivers or sewers or taken away by a sub-contractor for dumping (usually not in the East Midlands region). Whilst the disposal activity was necessary for the wellbeing of the firm, there were very large technical and organisational differences between firms which, to a great degree, determined the actual ratio of effluent outputs to their productive outputs. Quantitatively the major discharges in our sample came from dyers and finishers; however, in terms of impact on the environment, this group was not obviously the major polluter. Many of the firms we saw were discharging certain quantities of toxic oils and organic matter which may well have had a larger impact on the aquatic environment than the effluent from the dyers and finishers, and to a certain extent this was reflected in the way in which two firms in food processing and one in engineering made significant contributions to the local sewerage arrangements.

To place our results in context we begin first with some published data from the River Survey (5) and from the Severn Trent Water Authority (6). Following this, we then analyse our own findings which are split between the initial random sample and the later stratified sample of dyers and finishers which was made on the basis of the importance of the industry to the region's economy and also on the basis of its large contribution to the quantity of liquid effluent discharges.

In the Trent River Authority area, which covers most (but not all) of the East Midlands region, the percentage of industrial to total effluent is 43% (7) *. This is marginally less

* Including discharges to rivers.

than the national average, although if cooling water is included then the region has a substantially larger than average throughput. Generally the state of the water courses is better than the national average. The discharge of effluent is controlled by a good sewerage system; nevertheless in the southern part of the region there are some unsatisfactory storm overflows although their proportion is rather less than the national average.

Table 5.1 details the state and nature of industrial discharge within the overall Trent River Authority which included in its southern section a large industrial area not in the East Midlands. Total industrial effluent discharges excluding cooling water amount to about 10% of the national total. However, because of the very heavy predominance of electricity generation in the area, the Trent receives over 25% of all cooling water discharged nationally. The summer temperature of the river frequently reaches 27°C and, according to the Severn Trent River Authority (8), this high temperature is probably harmful to fisheries, enhancing the oxygen demands of the pollution load carried by the river and causing the depletion of dissolved oxygen. Any flow data for the East Midlands must, therefore, be corrected for the very substantial volumes from electricity generation. Since our survey is basically about small and medium sized firms which have no such large flows, we have emphasised in earlier sections the importance of textile dyeing and finishing in the region and within our sample, but in Table 5.1 this industry does not appear as a major discharger because the figures quoted include discharges direct to rivers and, although nationally textiles account for 2% of the volume of discharges, in the East Midlands most textile firms discharge to the sewerage in the first place. This was supported by evidence of our sample where we found only one textile finisher discharging directly to the river, and also by recourse to evidence of the Severn Trent Water Authority (9) which, out of a sample of 55 firms discharging into the river, noted only one textile finisher in the Trent catchment area. The majority of direct discharges were in iron and steel, engineering and food. The largest proportion of these industries is situated in the southern part of

TABLE 5.1 Major Dischargers in the Trent River Authority, 1971

	Million gals/Day
Chemical and allied trades	85
Iron and steel	40
Metal smelting and refining	27
Engineering	16
Coal mining	116
Paper	10
Food	9

the Trent River Authority and the East Midlands as an area suffers much less from pollution due to discharges direct to the river than the West Midlands.

Tables 5.2, 5.3 and 5.4 present an entirely different picture. These are data derived from the records of the Severn Trent Water Authority for the Nottingham area and relate to trade effluent discharges to the sewerage. The dominance of textile dyeing and finishing in this area is clear; in 1973 the industry accounted for in excess of 60% of all liquid wastes in all three size categories. This dominance declined in the medium and small/medium categories over the period 1968-73. The rate of growth of discharges in this industry was

TABLE 5.2 Volume of Trade Effluent from Small and Medium Sized Firms Employing 0–50 People in Nottingham C.B.C.
(all volumes in gallons) Index 1968 = 100 in parentheses

Nature of Industry	Apr 68 – Mar 69	Apr 69 – Mar 70	Apr 70 – Mar 71	Apr 71 – Mar 72	Apr 72 – Mar 73
Textile dyeing, finishing, etc.	32,820,000 (100)	38,623,000 (118)	36,791,000 (112)	34,518,000 (105)	39,110,000 (119)
Metal finishing/engineering	1,703,000 (100)	1,600,000 (94)	2,993,000 (163)	3,126,000 (184)	4,014,000 (236)
Chemical and pharmaceutical	866,000 (100)	814,000 (94)	750,000 (87)	698,000 (81)	825,000 (95)
Brewing (Maltings)	519,000 (100)	593,000 (114)	600,000 (116)	579,000 (112)	589,000 (113)
Dairying	633,000 (100)	577,000 (91)	471,000 (74)	540,000 (85)	673,000 (106)
Food Processing	14,980,000 (100)	10,414,000 (70)	9,437,000 (63)	8,663,000 (58)	5,882,000 (39)
Soft drinks	12,808,000 (100)	9,903,000 (77)	6,414,000 (50)	5,870,000 (46)	6,118,000 (48)
Printing	1,389,000 (100)	2,338,000 (168)	2,591,000 (187)	3,147,000 (227)	3,259,000 (235)
Garages	927,000 (100)	3,415,000 (368)	2,905,000 (313)	3,195,000 (345)	3,093,000 (334)
Miscellaneous	380,000 (100)	450,000 (118)	470,000 (124)	595,000 (157)	658,000 (173)
Totals	67,025,000	68,727,000	63,422,000	60,931,000	64,221,000
Number of firms	60	63	65	70	68

Source : Data supplied by Trent River Authority

TABLE 5.3 Volume of Trade Effluent from Small and Medium Sized Firms Employing 51-200 People in Nottingham C.B.C.
(all volumes in gallons) Index 1968 = 100 in parentheses

Nature of Industry	Apr 68 - Mar 69	Apr 69 - Mar 70	Apr 70 - Mar 71	Apr 71 - Mar 72	Apr 72 - Mar 73
Textile dyeing, finishing, etc.	251,348,000 (100)	261,208,000 (104)	250,018,000 (99)	260,074,000 (103)	276,219,000 (110)
Metal finishing/engineering	26,334,000 (100)	31,231,000 (119)	42,968,000 (163)	46,030,000 (175)	53,326,000 (202)
Tanning	24,675,000 (100)	27,270,000 (111)	32,318,000 (131)	49,528,000 (201)	42,766,000 (173)
Brewing	61,949,000 (100)	65,012,000 (105)	62,921,000 (102)	59,739,000 (96)	62,682,000 (101)
Dairying	22,228,000 (100)	24,764,000 (111)	34,362,000 (155)	39,286,000 (177)	55,874,000 (251)
Food	13,291,000 (100)	10,509,000 (79)	11,237,000 (85)	10,695,000 (80)	11,855,000 (89)
Soft Drinks	7,253,000 (100)	7,189,000 (99)	7,768,000 (107)	7,519,000 (104)	8,485,000 (117)
Printing	434,000 (100)	Min.charge -	Min.charge -	Min.charge -	Min.charge -
Garages	1,315,000 (100)	2,046,000 (156)	1,621,000 (123)	2,322,000 (177)	3,511,000 (267)
Miscellaneous	281,000 (100)	448,000 (159)	402,000 (143)	204,000 (73)	Min.charge -
Totals	409,108,000	429,677,000	443,615,000	475,397,000	514,718,000
Number of firms	34	34	36	37	37

Source : Data supplied by Trent River Authority

TABLE 5.4 Volume of Trade Effluent from Medium Sized Firms Employing 201-1000 People in Nottingham C.B.C.
(all volumes in gallons) Index 1968 = 100 in parentheses

Nature of Industry	Apr 68 - Mar 69	Apr 69 - Mar 70	Apr 70 - Mar 71	Apr 71 - Mar 72	Apr 72 - Mar 73
Textile dyeing, finishing etc.	972,357,000 (100)	996,261,000 (102)	1,001,118,000 (103)	914,781,000 (94)	931,347,000 (96)
Light engineering (mainly cooling water)	4,654,000 (100)	3,741,000 (80)	6,717,000 (144)	10,706,000 (230)	5,316,000 (114)
Chemical and pharmaceutical	71,103,000 (100)	70,708,000 (99)	114,958,000 (162)	86,050,000 (121)	88,703,000 (125)
Soap manufacture	667,000 (100)	53,218,000 (101)	49,792,000 (95)	50,741,000 (96)	44,955,000 (85)
Dairying	39,791,000 (100)	48,954,000 (123)	58,859,000 (148)	84,699,000 (213)	58,573,000 (147)
Food processing	12,855,000 (100)	11,406,000 (86)	10,769,000 (84)	9,443,000 (73)	10,514,000 (82)
Printing	7,768,000 (100)	12,888,000 (166)	10,636,000 (137)	9,529,000 (123)	10,985,000 (141)
Totals	1,161,195,000	1,197,176,000	1,252,849,000	1,165,949,000	1,150,393,000
Number of firms	24	23	24	24	23

Source : Data supplied by Trent River Authority

TABLE 5.5 Relationship between Number of Firms and Share of Total Effluent in Nottingham C.B.C.

	All Firms	Number of Employees			Total Effluent	Number of Employees		
		0–50 %	51–200 %	201–1000 %	(million gallons)	0–50 %	51–200 %	201–1000 %
1968/69	118	50	28	20	1,637	4	25	71
1969/70	120	52	28	19	1,696	4	25	71
1970/71	125	52	28	19	1,824	3	24	69
1971/72	131	53	28	18	1,702	4	28	68
1972/73	128	53	28	17	1,729	4	30	67

NB – Numbers do not necessarily sum to 100 due to rounding errors

Source : Data supplied by Trent River Authority

in fact substantially less than in other industries. This matches up with our sample evidence which suggests that, particularly in this industry with the onset of higher water and effluent charges, most firms are making efforts to curtail effluent by recycling water or using less of of it by adopting less water-intensive processes. This seems to have occurred most in the medium sized category; possibly because really small firms do not always have the scientific and financial resources to make rapid changes in the face of stiffer pollution controls. The same effect may also be observed in the brewing and food industries but in metal finishing, light engineering and garages the growth of effluent has probably kept up with growth of output. This seems quite consistent with the hypothesis that in these industries there have been substantial changes in process technology and that, to an extent, pollution control technology lags behind process technology. The necessary impetus in the form of effluent charges and controls itself is lagged and the main pressures are for greater productive efficiency rather than effluent efficiency in the initial stages of development.

More important still in the present context is the relative unimportance of the small firm as a discharger of effluent. Table 5.5 illustrates this; in 1973 17% of the firms discharging to the sewers accounted for 67% of total effluent, the remaining firms (which were in the same category as our sample) accounted for only 34% of total effluent whilst the very small firms had an almost negligible impact on total discharges. If the large firms employing more than 1000 people had been included in the evidence, the small firms would seem even more unimportant. This evidence is supported by data from the Derby County Borough which shows the smallest 50% of firms have only 12% of the effluent and the smallest 75% have only 40%. It was not possible to make an exact comparison with the data from Leicester but the data do suggest a similar pattern. Thus we conclude that the role of the small and small/medium sized firms as water polluters is relatively minor. However, this by no means suggests that the control cost to the firm is small, nor is it an entirely general one since, in some instances, (such as the food industry) a substantial proportion of pollution is indeed accounted for by the very small sized firms. The opposite polar case to this is textile finishing, where small firms are really unimportant polluters.

Control and Charges

The levying and extent of effluent charges before local authority reorganisation varied substantially between local authorities. Generally some hybrid of the Mogden formula (10) was used which took account of BOD *, COD**, conveyance, volumetric sludge factors and a charge for suspended solids.

As we have indicated at the start of this chapter, there are strict limitations on the discharge of toxic materials like cyanide and arsenic and also factors such as the pH of the effluent and its temperature. The two most striking features of the system were the enormous variability in charges and thus their differential impact on firms located in different areas and the overall lack of policing and prosecutions that there has been. In Leicester, for instance, only one firm has been prosecuted. This firm was operating a plating shop and had been consistently contravening discharge consent conditions with respect to pH, cyanide and toxic metal content. The firm was found guilty, yet was fined only £25 plus £11 costs. In Nottingham the only prosecution had been against the Bulwell

---

* Biological Oxygen Demand

** Chemical Oxygen Demand

Finishing Company. The firm claimed exemption from payment of trade effluent charges under a notice of direction served by the Corporation under sections 55 and 57 of the Public Health Act 1961. The company had an agreement with the Corporation made under seal in 1908 and claimed that this agreement was preserved by section 7(4) of the Public Health (Drainage of Trade Premises) Act 1937 and that the Corporation were therefore not entitled to serve the notice of direction. Judgment was given in favour of the company. Possibly the paucity of prosecutions and a level of effluent charges well below the national average are evidence of the political nature of the pricing of effluent in the East Midlands. Certainly this would conform with our evidence concerning air pollution and the different imposition of regulations in that area. The small firm, in particular, is at a distinct advantage because of difficulty of detection. According to R. Fearn [11], Divisional Controller of the Severn Trent Water Authority "... many small firms are unaware of the (pollution) legislation and have to be informed of their duty to serve a notice of effluent discharge two months prior to commencement..." "... many firms are discharging illegally, it is quite impossible to say how many such firms exist within the drainage area...".

Certainly our survey confirmed this; numerous firms admitted to occasionally tipping toxic material down the drain and most seemed to agree that there was usually some prior indication of a visit from an inspector. With the reorganisation of local authorities, effluent charges were initially increased five-fold but, after protests from various industry associations, it was agreed to spread this increase over a period of five years. Since concurrently with this water charges have also been raised some firms, particularly in the textile industry, seem bound to suffer. Those firms we interviewed were very conscious of the problem and were actively investigating possible ways of reducing their impact. Furthermore, since within the price rises there were moves towards the equalisation of charges, this would particularly hit the urban based textile finishers in the East Midlands located in and around Nottingham and Leicester. The Knitted Textile Dyers Federation told us that the average textile factory was paying £3,600 per annum in effluent charges prior to the increases and that, if the rises were absorbed in one year, effluent charges would cost about £18,000. Such a figure would represent 2% of turnover on average [12].

Replies from the Random Sample

Our questionnaire conducted on a case study basis was designed to investigate how individual firms dealt with their pollution and, in the case of liquid effluent, what impact this had on the firms' operations in the short and long run. To a certain extent our knowledge of firms and industrial structure would suggest that, in the short run, the firms might adopt a variety of measures to deal with cash flow and liquidity problems. In general these would involve changing input/output ratios within relatively narrow parameters, price or employment changes and ultimately ceasing to trade. In the long run we might expect substantial changes in processes and products accompanied by movements in capital: labour: output ratios. In this section we deal with the responses of the firms from our initial random sample whilst in the next section we deal with the stratified sample of dyers and finishers. In response to our first question asking firms if they discharged any liquid wastes, twenty out of a total of fifty-eight replied positively. As expected, all textile dyers and finishers had a liquid waste problem and the majority of the other discharges of trade effluent were from engineering and metal manufacturing; although the largest discharge came from a firm in the food processing industry, discharging 27,000 gals/day. The nature of the liquid waste discharged is partially covered by Table 5.6.

TABLE 5.6 Nature of Liquid Wastes Discharged

	Number of Replies
Acids	10
Oils	8
Other	10

NB : Figures add to more than twenty due to multiple responses.

There was a poor response to this question because sometimes firms wanted to maintain secrecy over the type of process used or, more often, they felt that their wastes were not adequately covered by the classifications. Clearly the basic split was for dyeing companies to discharge acids, alkalis, suspended solids and some colourants, and for firms in the engineering industry to be heavy dischargers of oil. Almost without exception dyeing companies we interviewed were not discharging metals, as had once been the case when metal-based dyes were the norm. In the random sample almost half of the dyers were using solvent cleaning although solvent dyeing is still in its infancy. The use of non-aqueous cleaning processes was generally confined to the larger companies in the sample and, at the time of the interview, could not be correllated with the cost of water or effluent charges and was generally used because of specific product quality or fabric considerations. Generally, discharge of solvents was limited since there is a constant recycling of contaminated material back to the supplier. This was also the case with oils discharged, although most of the lubricating oils used were soluble in water. The metal-finishing industry used substantial quantities of acid for cleaning and for electroplating but the only acids discharged to the drains were the very weak concentrations resulting from rinsing after cleaning. The other important liquid effluents were : cooling water (4 cases, 2 in engineering and 2 in metal finishing), dilute animal food waste (2 cases, both resulting from cleaning processes), trace nitrates in solution of corrosion inhibitor (engineering), dilute ink (1 case, printing trade) and sodium sulphate and lime in dilute solution (1 case, food trade). Those firms having no liquid waste problems were in brick making, food manufacture (2 cases), refractory manufacture, saw mills and engineering (15 cases). Our impression was that most firms discharging small quantities of any liquid wastes other than non-soluble oil could, without too much difficulty, discharge direct to the foul sewer without detection. One firm which had been stopped by the local authority from dumping its caustic sludge even on approved dumping sites, frequently threw it down the drains in small quantities without detection. The policing by local authorities of consent conditions varied because of differences in size of firms, proximity of firms to each other, the hetrogeneity of firms in an area and output of domestic sewage. The one area where detection seemed easier was with non-soluble oils; in two cases of accidental discharge the firms responsible had been traced. We have little evidence that toxic materials go illegally down the drain, but it was clearly impossible to get this sort of information from respondents. It is not a difficult task to discharge illegally in small quantities.

The wastes of the major water polluters (by volume, dyers and food) were biodegradable whereas many of the effluents discharged by engineering firms and those in metal manufacture were not. The most surprising aspect of answers to this question was that a large proportion (about 25%) of respondents in all trades were not aware of the meaning of "biodegradable";

this was particularly so with the very small firms, even in dyeing and finishing, where the use of biodegradable dyes was almost complete. To a certain extent the use of biodegradable inputs could be linked with economic progress; for example there had been a movement towards the use of soft detergents, although it is as well to note that in the early beginnings of the textile trade all detergents were soft. There was then a movement towards hard detergents and now back to soft again. This is, of course, quite consistent with our model of pollution in Chapter 1, since it is likely to be the result of changing levels of satiation (minimum and maximum) and changing technology (today's soft detergents are better than the very early pure soaps).

The use of biodegradable materials - which may at first sight seem to be beneficial since they do not remain permanently in the environment - paradoxically has the effect of removing oxygen from the water but, of course, these materials are broken down in treatment at the sewage works and are therefore not passed on into the water course. Non-biodegradable materials can only be removed at the sewage works with great difficulty and are therefore more likely to remain permanently in the environment. Thus, unless effluents containing biodegradable materials are treated on site or at the sewage works, the use of such materials would lead to a reduction in the dissolved oxygen content of water courses.

Naturally no two materials behave in precisely the same way; although both may be classified as "biodegradable", it is therefore possible that of two "biodegradable" materials one has a low five-day biological oxygen demand but a high twenty-day biological oxygen demand and will therefore de-oxygenate further upstream than a substance with a high five-day oxygen demand. In such a situation, it is by no means clear which of these two materials would be best to use.

In Lake Eyrie, wood pulp wastes have stimulated plant growth leading to a further increase in oxygen demand and hence further eutrification of the water. However, no firm that we contacted was large enough to have had such an impact on the eco-structure, although two food manufacturers potentially could have filled this role and, in anticipation of this, the local authority demanded that these firms make a significant contribution to the cost of the sewerage system.

Only six firms did not use the sewerage to dispose of any of their effluent. Three of these firms discharged to their own water course; these were two dyers and finishers and one firm in metal manufacture. In each of these cases the effluent was treated before being discharged to the river and it was the location of the firm that prevented it from using the sewerage. Normally the firm had been given the option of using the sewerage system if it had been prepared to pay the costs of connection from factory to main sewer. None of these firms was small; all had in excess of 200 employees. However, even for large quantities of effluent it was still much cheaper to build the firm's own treatment plant than pay connection charges (in one case one-fifth of that cost). In all instances there were very considerable economies of scale either for linking to the sewerage or for building the firm's own effluent treatment plant. These economies arose either through the large fixed cost element of building a pipeline or through the simple rule that the cost of an effluent treatment plant (normally based on a large settling tank) increases by the square but the volume increases by the cube. The substantial costs and large scale economies involved for a firm not located on the main sewerage system probably explain to a large extent the dearth of small firms discharging direct to water courses and the decline of water-dependent industries in rural areas.

There were also three firms which used only private contractors to get rid of their liquid effluents; these were all firms in engineering and metal finishing where the wastes were highly toxic or had some potential value through recycling. In addition to this, four firms used private contractors on occasions to get rid of their toxic or reclaimable wastes; three of these were in engineering and one was in the food trade. Generally, removal by private contractor exhibited substantial economies of scale because of the existence of a minimum fixed charge and our consultations with the major waste removing firms confirmed this, although as a factor militating against really small firms this was nothing like as substantial as those involved with building a treatment plant. The disposal costs for certain wastes could be substantial because local authorities would often not allow even small quantities to be tipped in their area; we found that in several cases the wastes had to be taken to Essex or, in one case, sealed in drums and dumped in the North Sea. These high costs (especially for the very small firms) and the lack of adequate policing facilities have led to the establishment of a number of unethical sub-contractors who are probably fly tipping. If the environment is to be protected from possible disastrous consequences, there may be a genuine economic case here for local authority help. At the time of writing this report, few local authorities had industrial waste disposal transport and none dealt with toxic substances in any way.

Because only a small number of firms discharged direct to the rivers we found only five instances where the waste was pretreated before disposal. Usually this pretreatment involved substantial costs; in one case the capital cost of the process was £40,000 and running costs amounted to almost £5,000 per annum - a not inconsiderable sum given that the turnover of this firm was only £250,000 per year and that it had had to install this treatment plant when wage and raw material costs of production were rising rapidly and when prices were subject to government control. Treatment plant was usually used to settle out solids, neutralise acids or alkalis, or, in two cases, to cool the effluent to an acceptable temperature. With the revised water charges, effluent charges and increased cost of raw materials, more firms were considering some form of treatment of wastes. Generally any moves in this direction were because of the possible benefits of heat recovery or, in a few cases, the recovery of chemicals. Even though effluent charges had been increased substantially, the representatives of the firms we interviewed saw the main benefits of treatment arising from recycling rather than effluent charge avoidance. The other alternative is to change processes and, in dyeing and finishing, there was considerable evidence of this. Whilst treatment plants favoured large scale production, effluent charges were not biased in favour of large or small firms. There were often substantial benefits from the treatment of waste but in the firms in our sample this technology was largely ignored and within the organisation there did not seem anyone with responsibility for this aspect of production and consequently it was frequently ignored. There were some exceptions to this; in our case studies we refer to a dyer and finisher that derived very substantial production benefits following investigation of treatment processes of his effluent. In another case in the food industry a firm was able to make a net profit of £2000 per annum after instituting a pretreatment process for its wastes. In both these cases the pressure on the firm was not in the form of effluent charges but rather because the local authority refused permission for the discharge of certain substances. Rigid controls seemed in these instances to act as catalysts for productive and profitable technological innovation although the smallest firms are unlikely to be able to innovate quite as effectively as the medium sized firms that employ qualified scientists and engineers.

The total cost to the firm of effluent disposal consists largely of the trade effluent charge levied on it. Table 5.7 shows estimates of such costs for 25 members of the Knitted Textile Dyers Federation in the East Midlands region and Table 5.8 shows how they were expected to change by 1975. The increase has been generally somewhere between 2½ and 4½ times, but in exceptional cases the increase has been outside these limits. Table 5.9 gives the new scale of charges throughout the region and shows how the variability throughout the region is diminishing.

TABLE 5.7 Effluent Charges for a Sample of Dyers and Finishers, October, 1974

Code No.	Period Ended	Suspended Solids	C O D	Annual Discharge	Annual Cost
		ppm	ppm	Th.Gals	£
1/L	June 74	176		57,480	3,161
2/L	Dec 73	29		53,770	1,653
3/N	June 74	55	465	59,000	2,537
4/N		63.5	653		
5/L	Sept 74	18	70	15,500	589
6/N	June 74	60	1130	30,800	2,125
7/L	June 74	57	510		
8/L	June 74	44	880	27,447	1,128
9/O	Mar 74	95	482	25,392	2,226
10/N	June 74	249	895	55,938	3,606
11/L	Mar 74			24,641	1,355
12/L	June 74	50	550	64,220	2,055
13/L	Mar 74			16,256	892
14/N	Mar 74	49	540	83,000	4,977
15/N	Dec 73	50	490	101,040	6,000
16/L	Sept 74	20	520	9,600	520
17/N	Sept 74	106	1163	21,753	1,626
18/L	Sept 74	17	50	5,908	295
19/O	Mar 74	127	(BOD) 771	27,578	1,403
20/O	Mar 74	63	1461	27,500	3,875
21/O(a)				38,400	
22/O	June 74	45	680	81,000	3,798
23/N(b)	1969	133	654	40,000	Nil
24/N	June 74	45	46	34,882	1,469
25/O	Mar 74	106		8,240	561

L = Leicester firms N = Nottingham firms O = Other firms

(a) No charge has been made by local authority who consider that discharge from dyehouse has beneficial flushing effect.

(b) Exempt from charges by virtue of old agreement.

COD = Chemical Oxygen Demand BOD = Biological Oxygen Demand

Source : Knitted Textile Dyers Federation

TABLE 5.8 Effluent Charges for 1975/6 for a Sample of Firms shown in Table 5.7

Code No.	Charges (pence per gallon)	Annual Cost £	Increase %
5	6.5	1007	71
7	11.1	-	76
8	14.6	4007	256
12	11.4	7301	256
16	10.8	1037	100
18	6.5	384	30

Source : As for Table 5.7

TABLE 5.9 Average Estimated * Water Authority Charges for Dyers and Finishers in the East Midlands (1975/76)

	Pence/thousand gallons
Nuneaton	44.3
Leicester	48.9
North West Leicestershire	54.7
Sherwood	41.2
Nottingham	32.3
North Derbyshire	49.1
South Derbyshire	41.5

* Estimates based on Trade Association data of its members' COD, BOD and SS.

On average the dyers and finishers in the random sample were discharging approximately 20 million gallons of water per year which would amount to almost £10,000 in effluent charges; some were discharging much more and one firm in the food industry was discharging almost 100 million gallons per year. Effluent charges alone at today's prices would amount to almost £50,000 per annum whilst, in addition to this, the firm employed several men to look after the system and had already made a contribution of the order of £50,000 towards the cost of installing additional capacity at the local sewage works. The annualised costs were in this instance approaching £100,000 per annum. Many firms, of course, had very low or negligible effluent charges and for not inconsiderable discharge were only paying in the region of £1000–£2000 per annum. Indeed, some companies did not perceive effluent charges as large. Much worse was any contribution to a sewerage scheme or new drains. One firm which had just spent £18,000 on new fume arrestment equipment found it had also to contribute £7000 to the cost of the drains. Over a long period of time these costs could be considered very minor but, in the short run when a firm is faced with a cash flow crisis and when bank lending is tight (the sort of conditions that existed in 1974/5), then these sort of costs are those that can cause bankruptcy and close down a business. For instance, this happened to the Sherwood Dyeing and Bleaching Company; we interviewed their works manager a few days before closedown and he was in no doubt that the combined effects of pollution control costs for air and liquid wastes needed over a short period of time were partially responsible for the firm's closedown. Another firm which still operates in the Nottingham area claimed that this is why it had closed one of its older factories. In no way are we suggesting that these controls are unfair or unjustified but it must be recognised that even though a cost may be small on an annualised basis it can cause severe cash flow problems in the short run, especially for the smaller firm whose sources of finance are less secure.

We came across few firms with really high concentrations of toxic materials and, where acids and alkalis were used, firms often used settling tanks for one to neutralise the other and so, apart from those few using private contractors for waste disposal, the data in Tables 5.7 and 5.8 are probably fair guides to effluent charges. Our experience suggests that in most cases where firms do not use the sewerage for disposal these figures can be doubled to give a final total cost. Effluent charges (and also water charges) have been some way below true cost up to 1975 and, when the charges were raised, there seemed some evidence that the rise helped initiate savings which involved zero loss to production. Only three firms we talked to felt the charges were too costly; most respondents, expressing an individual as opposed to a corporate view, thought that in general they represented good value.

Accidental discharges of effluent to the sewerage or water course were possible, although in only 10 instances did firms admit that this could happen. The majority of these firms were in engineering and metal finishing where carboys of acid could leak and seep down the drains. Several firms admitted that this had happened, although only one had ever been prosecuted. This evidence seems to tie in with our doubts about the policing of water polluters. Finally, whilst most firms sampled their effluent from time to time, only ten had any form of automatic monitoring system and these were mainly in the textile trade. This may well represent a way by which the water system could be policed at a low cost in future, since those firms using automatic monitoring found it relatively simple to use.

Generally there were few dangerous water polluters in our random sample and, to a certain extent, this is probably a reflection of the regional picture where volume of water pollution is more important than type. There were some substantial problems for small firms, particularly in dyeing and finishing, and it is our stratified sample in this trade to which we now turn.

Textile Dyeing and Finishing

As we have said previously, this industry is an important trade in the East Midlands and also a significant polluter of the environment in volumetric terms, although the toxicity of its effluent is probably slight. The structural data on this group have already been referred to in Tables 4.18 - 4.20. There are very substantial variations in quantity and type of both inputs and effluent output, depending on the specific nature of the fabrics to be dyed and the processes used, thus since the wastes are so varied, monitoring cannot be routine and the collection of information is made more difficult. The recorded values for B O D and pH of the effluent are extremely wide and there is also substantial variation in effluent temperature. Some of the basic pollutants can neutralise each other, although the discharges of these occur sequentially and not simultaneously and thus some form of settling/mixing tank is necessary in order to minimise the pollution load to the sewerage works.

All firms discharged some liquid wastes and, even though solvent cleaning and dyeing were accepted innovations in some parts of the trade, their use was not widespread; the traditional methods incurring heavy water usage were still dominant. Water usage was often the basis of the original company location and water costs up until recently had been considered as basically a rigidly given input to the business operation which need not be altered to any large extent. However, with the recent 1974/5 increases in water charges referred to previously, firms had become concerned about water usage. About 25% of our sample were actively investigating water reclamation; this could generally be brought about by missing out a final rinse or re-using rinse water in the dyeing operation. Scouring and dyeing simultaneously would also reduce water input but this might affect the final quality of the product by reducing colour fastness or making some marginal difference to the finish, whilst installation of more sophisticated water re-use equipment results in a high level of capital expenditure that many small firms cannot afford without help. Water usage and the structural characteristics of the firms were, in fact, quite closely related; the two structural factors (i.e. excluding technical and end-product considerations) which seemed to affect the level of water usage and consideration of new water reducing innovations, were the size of the firm and the degree to which the firm belonged to a larger integrated group. Thus the three largest firms in our sample had a rate of water usage per employee less than 50% of the average. A finer breakdown is not available because some firms would not give us enough detailed information on numbers employed. Table 5.10 illustrates how the degree of integration seems to be negatively associated with water usage.

TABLE 5.10 Water Usage and Vertical Integration *

	Rate of Water Usage		
Number of firms in	High	Medium	Low
(a) Integrated groups	0	2	6
(b) Independents	5	3	2

* Water usage was based on the ratio of the number of employees/ the quantity of water in thousands of gallons used per week. Low rates of water usage were 10-30, medium rates 30-60, high rates 60+.

Further questions soon revealed that the firms in the integrated groups had already begun serious work on water savings, adopting various new dyeing processes using solvents or a variety of water re-use systems. Scouring and dyeing simultaneously reduced water usage but needed extra additives and auxiliary agents, thus the quantity was reduced but the quality of the effluent was not always improved. Other methods used, like solvent cleaning and dyeing, used no water whereas high temperature (HT) dyeing halved the liquor : water ratio. Dispersed dyes could also be used and needed fewer rinses for a given degree of fastness; however this particular innovation resulted in a higher BOD and could only be used with terylene. Foam dyeing can be used for hose and has a much lower rate of water use. Until the year in which we questioned the firms, there had been so little concern with water usage and effluent charges that only one firm had paid much attention to its water inputs and only three had continuous monitoring of outputs. The changes in charges had obviously stimulated search for new technology or operating systems in this area. The implications for volume of inputs was quite clear but the change in nature of outputs was less certain. Many independent firms had neither the financial resources nor the technical know-how to get to grips with the problem of water conservation and this failure ultimately manifests itself again in the form of higher effluent charges for the non-innovating firms. This was one indication that an industry with a relatively well-known and easy-to-understand technology was rapidly becoming more complex and would begin to favour the larger, more integrated firm. From the point of view of pollution control and the ability of the industry to pass on the costs, this meant that in the short run we could expect, and indeed found, many firms leaving the industry whereas in the long run the increasingly complex technology would lead to higher entry barriers which in turn would to an extent protect the firms already in the industry from competition and might facilitate the passing on of costs. Thus in the short run there would be little or no effect on prices whereas in the long run the impact on prices may be expected to be greater. There were a large number of other technical factors impinging on water usage which would explain some of the variance of our data, but one final factor deserves mention and that is the variability in the throughput of the machines; generally the firm had to use a given amount of water whether its machine had a full load or not and thus the larger firm through the law of large numbers would always be in a better position to prevent partial utilisation of machines. Also because of smaller batch size and greater variability of throughput in the small firm, the large firm may be able to plan water usage more efficiently.

A further problem arises through space constraints; many of the smaller, older firms had located in an urban area where space was now severely limited and this was to represent a considerable constraint for any firm contemplating water re-use or some form of pollution control treatment, since such plant tends to be space intensive.

In all cases our respondents used biodegradable dyes, although in four instances chrome dyes were used from time to time. Generally there seemed almost quasi-integration between the dyestuff manufacturers and their users, in as much as the users could sometimes use the dyestuff manufacturers' expertise to help them solve production or pollution problems.

The quantities of effluent discharged per week where given are in Table 5.11. BOD, suspended solids and pH value varied substantially between firms depending on what they were dyeing, thus only volume data can be aggregated in any meaningful way. We have no evidence that particularly toxic materials were being discharged to the sewerage (with the exception of occasional small quantities of metal dyes). Although it was not clear in some cases how solvents were disposed of, the individual problems that were encountered are best described on a case study basis (see Chapter VIII).

Generally the volume of effluent was positively associated with size of firm but this relationship was not linear and showed a decreasing rate of discharge over the total size range, emphasising the very substantial economies of scale in water treatment. As we have already said, the type of effluent varied with type of fabric and there was some evidence to suggest that, using old technology synthetics resulted in more effluent than natural fibres but new processes like the use of dispersed dyes which had been developed for dyeing terylene and transfer printing for synthetics in general will lead to lower volumes of pollution. One problem that we noticed in our random sample and which also recurred in this sample, was the apparently strong relationship between quality of end-product and quantity of effluent. Several respondents blamed the very high quality demands of chain stores buyers for an increase of effluent per unit fabric dyed. We also found evidence to

TABLE 5.11 Volume of Liquid Discharged by Textile Dyers and Finishers

Volume of Discharge 1000 gallons per week	Number of Firms
Less than 50	6
50-249	4
250-499	4
500-999	4
1000+	6
	24 *

* Not all firms were able to give us accurate data on volumes discharged.

support our earlier belief that, because of the cheapness of water and effluent charges, many firms just had not tried to reduce the quantity and nature of effluents; however, with increased water and effluent charges, a remarkable number of firms had made small savings in effluent output merely by realigning existing inputs more efficiently.

Fourteen firms out of the total stratified sample treated their effluent in some way. Nine used a settling tank, five used a heat exchanger and only one had a full treatment facility because it was discharging into the river. Although some of our respondents claimed that firms operating in other parts of the East Midlands had less demanding consent conditions, we found no evidence to suggest that existence of treatment plant was in any way related to geographic locations, with the one exception of that firm without mains drainage that had to incurr very substantial expenditure either to be connected to the main sewer or to build its own treatment plant. The use of heat exchangers seemed in most cases to be a pollution control device that paid for itself through recycled heat and energy. Some of these were a very simple design and not needed for the firm to comply with consent conditions. Generally the settling tanks were simply a means for the acid and alkaline wastes to neutralise each other, although sometimes additional chemicals were needed to ensure neutrality of the effluent. There were also in-process changes that might lead to reductions in effluent outputs such as jet dyeing, HT dyeing and continuous dyeing. We have dealt with most of these already under the auspices of water re-use and conservation and thus their net cost to the system might well be negative since they lead to water savings and quality changes. Generally where there was an in-process change or end-of-line control (like a settling tank) this was something from which economies of scale could be derived, although not to any very large extent because of complex technology but merely because the cost of building simple treatment plants like settling tanks increase by the square whereas their capacity increases by the cube. Unfortunately cost data of control equipment was not always available and was sometimes unreliable when it was quoted, but the cost of a settling tank would be rarely greater than £15,000 whereas the cost of full treatment plant had an initial capital cost of £40,000, a replacement cost of £70,000 and a life of 10 years, with running and maintenance costs being about £1000 per annum. Effluent charges have been well documented in the previous section and this sample largely conform to that previous evidence; Table 5.12 illustrates.

These data relate to the pre-1975 increases in charges and so some may have increased threefold whilst others increased a little less than this. Even these increases do not make the charges enormous and whilst it is impossible to generalise, we would estimate that in

TABLE 5.12 Effluent Charges pre-1975 Increases

Firm Code	Location	Volume of Effluent (000 gals/week)	Size of Firm	Charges (£ per year)	Pretreatment
1	L	500	Small	2,500	No
2	O	N/A	Large	No charge	Yes
3	N	1,250	Large	2,500	No
4	N	500	Medium	4,800	Yes
5	N	800	Large	2,400	No
6	O	N/A	Medium	3,500	No
7	N	N/A	Medium	3,000	Yes
8	O	4,000	Large	20,000	Yes
9	O	1,000	Large	8,750	Yes
10	N	500	Large	8,000	Yes
11	O	N/A	Medium	No charge	Yes
12	N	350	Medium	1,300	No
13	N	250	Medium	1,500	Yes
14	N	50	Small	1,000	No
15	N	1,400	Medium	18,000	No
16	O	1,500	Medium	1,600	Yes
17	L	1,200	Medium	3,300	No

L = Leicester
N = Nottingham
O = Other

Small = less than 100 employees
Medium = 100 - 500
Large = 500+

the post-1975 situation, the charges would represent only about 5% of turnover given the most pessimistic of conditions; although total liquid, solid and air pollution control costs may treble this estimate.

The problem then would not affect all firms equally but it would have its worst impact when a combination of circumstances such as increased effluent charges, stenter equipment and settling tanks were all required at the same time, particularly since this industry has a notoriously low level of profit. Under these circumstances only large or integrated firms could hope to survive.

Some additional points can be made about the Table :

i. Whilst the volume of effluent varies with the charges, this is clearly not a proportional relationship and thus there is considerable variability which emphasises that non-discriminatory measures to either control pollution or to help firms pay for pollution control will not be successful since pollution and its control is often specific to individual circumstances.

ii. There is also no evidence to suggest that where there is pretreatment there will be a much lower level of charges, although our data cannot tell us whether pretreatment of effluent has prevented higher penal charges.

iii. It was suggested to us that different geographic areas paid vastly different charges; this may be so, but our evidence cannot support it at present.

iv. The evidence supports the view we have made elsewhere that the existence of pretreatment facilities is normally associated with large size; replies to our question asking why there was no pretreatment would give added weight to this direction as over 90% of the replies gave excessive cost as the main reason for not installing treatment, although the majority of these firms felt that in the near future they would be forced into additional measures or substantially increased sewerage charges.

Perhaps more important than cost was the fact that 19 of the firms interviewed felt that lack of space was a constraining factor affecting the potential installation of treatment facilities. This represented 60% of total replies and is a factor we consider to be highly significant, albeit a factor specific to industries like textile dyeing and finishing which are traditional and have tended to be sited in urban areas.

We also talked with various manufacturers of plant and materials and we found that :

(a) ICI reported to us that they no longer sell sulphur dyes which require the addition of sodium sulphide to reduce them. We have also learnt that the oxidisation process which previously used chrome has now been changed to a more costly and difficult process using perborate. Sulphur dyeing is something normally performed by smaller companies with unsophisticated equipment.

(b) ICI also report that sales of solachrome dyes are being discouraged as they need bichromate to fix the shade; sales of direct dyes needing metallised after-treatment have been reduced as have sales of vat dyes which require sodium hydrosulphide in the dye bath.

(c) The use of sulphuric acid to equalise effluent is now being discouraged by river authorities as there are restrictions on sulphates. Hydrochloric acid, a much more expensive substitute, is having to be used instead.

(d) ICI report that the use of carriers (phenolic compounds) in polyester dyeing is being replaced by high temperature dyeing (which does not require the use of a carrier) but which does lead to high temperature effluents.

(e) The Mortenson System of water re-use (reported in the International Dyer, June 4th, 1971) is a method of cleaning waste dye liquor and re-using the water. The process involves the addition of various chemicals to an effluent stream and then passing the liquor through a sand filter, a cationic-exchanger and an anionic exchanger. Initial tests suggest the cost of the recycled water to be in the region of 20p per thousand gallons. A figure rapidly becoming economic, although none of our respondents had used it. An alternative to this is the Rotapur system which essentially involves two stages :

 1. the separation of solid matter by chemical filtration and flotation, and
 2. the treatment of the clarified water by a catalytic oxidation method.

For a plant using 5 M^3/hour the capital cost would be of the order of £12,000 with an operating cost of 3.85p per M^3 (17.5p per 1000 gallons) (1974 prices).

In addition to these there have been substantial changes in dyeing techniques which have either concentrated on one of three main areas :

1. high temperature to reduce the need for carriers;
2. solvent cleaning and dyeing, completely obviating the need for water; and
3. continuous bleaching/dyeing using the principle of linked processes and production planning to save water and heat.

Finally we should reiterate the claim of one of our respondents whose firm was a member of the Courtaulds Group; he said that the available engineering expertise provided by the group was an invaluable asset when dealing with pollution control, water usage and major process changes. This is another factor supporting the view that the small independent manufacturer is at a substantial disadvantage when faced with environmental control measures. Indeed, our overall conclusions to this section relate to the changing nature and technology of the industry, the technology has become more complex, foreign competition has increased enormously and it is our view that pollution legislation and charges are just another factor which will militate against the small firm in an industry which was once dominated by the small family business. There are now very substantial benefits from large scale and pollution control is just one of these.

Ferrous Founders

None of the very small ferrous founders we visited had a liquid effluent problem but five of the 14 firms using wet arresters on their cupolas revealed that they had to pay effluent charges and/or incur some pretreatment costs. A BCIRA Broadsheet [13] describes how the absorption of sulphur dioxide in wet arresters leads to the formation of sulphuric acid which has a highly corrosive effect on the arrestment equipment. The water in these arrestment plants is continuously recycled, with solids settling out as a sludge in a settling tank where any necessary make-up water is added. Periodically the settling tank needs to be cleaned out and the sludges disposed of; by the addition of soda ash to the process water, the water can be neutralised. One of the firms operating a plant of this type told us that the water treatment costs were £2.50 per day (say £700 per annum) and that disposal of the sludge - which can be very viscous and is not easily handled - cost £50 per week (say £2500 per annum). In most of these cases, although firms were occasionally discharging liquid wastes, the effluent charges were negligible. One firm (which is referred to in Case 21) was discharging its filtered waste into a nearby stream, which on the admission of management was causing considerable pollution. It had not, however, been asked by the river authority to clean up its discharge. We feel that local authorities have in the past viewed founders quite rightly as mainly air polluters and have thus ignored their liquid wastes. This perception bias seems to have continued even when some founders had in fact become considerable water polluters. The firms themselves seemed more aware than the controlling authorities and two firms in our sample had automatic monitoring equipment on their discharge.

Two of these five firms had, because of the particular type of work they were undertaking, large scale water pollution problems and both, in fact, were discharging direct to rivers as well as to the sewerage. One was discharging small quantities of acids and oils in addition to its sludge from the dust extraction plant and the other was discharging process water from concrete manufacture. Both firms had some sort of effluent pretreatment; in the first the river authority had asked for a water cooling plant to be installed to ensure an appropriate temperature of discharge (the sludges were simply flushed down the sewerage with the addition of large volumes of water). The other firm was neutralising the causticity of its waste by bubbling carbon dioxide through it; if this had not been done the firm would not have been able to meet the consent conditions laid down by the river authority. In addition to an annual charge of just over £500 for its clean water, this firm estimated that liquid effluent disposal cost it £6000 per annum, taking into account the running costs and capital charges of the neutralisation plant and the settling pond needed for the gas cleaning plant.

All these plants thought that accidental effluent discharge was possible but none had been prosecuted for any offence. In addition to the continuous monitoring of effluent already mentioned, one firm had erected barriers to guard against oil pollution - a matter not raised by other respondents.

We therefore conclude that liquid effluent disposal problems for most ferrous founders are negligible but the installation of wet gas cleaning - which inevitably will become more and more prevalent amongst the smaller firms - brings with it the need for some pretreatment facility, at least in the form of a settling tank, the operating and capital costs for which are quite low.

The Local Authorities

Responsibility for water supply and reclamation was removed from the local authorities and vested in the new regional water authorities on 1st April 1974. It is therefore appropriate to consider how much of the new water authorities' resources are devoted to industrial pollution control. The Severn Trent Authority shows that on average 520 million gallons of effluent are treated every day, of which 101 million gallons (or 19%) were trade effluent. Of course the proportion of trade effluent varies from area to area; in the mid-Trent division trade effluent accounts for about 13% of the total flow, although at one works alone trade effluent accounts for 30% of the flow. There is no direct relationship between trade effluent flow and the proportion of reclamation costs but, on average, treatment costs tend to be higher for trade effluent than domestic sewage and we could therefore estimate that, in the mid-Trent division of the Severn Trent Authority, treatment of trade effluent accounts for approximately 15% of total reclamation costs. In this area a total of approximately 650 establishments come under trade effluent control and it is also necessary to add in the costs of laboratory analysis of samples, inspection of sites and other aspects of negotiations and control of traders; these items would add just over 1% more to the cost of disposal of trade effluent. In addition there are certain other costs that are non-identifiable and for which allowance has not been possible and in other areas must be added costs of investigations into the quality of discharges to rivers as well as to the sewerage. We would therefore conclude that reclamation of trade effluent, which comes from a comparatively small number of establishments, accounts for approximately one fifth of total reclamation costs.

Conclusions

We have considered liquid effluent disposal problems in our original sample of firms and in our specific samples of textile dyers and finishers and of founders. We conclude that the role of the small and medium sized firms in water pollution is comparatively minor except in a few specific trades such as foodstuff manufacture. Even in textile dyeing and finishing the small firm plays a minor role. Most firms were discharging to the sewerage although a few were discharging to rivers; the installation of pretreatment plant was almost universally limited to the latter plants. Few firms monitor their effluent automatically and very few have been prosecuted for water pollution. The costs of pretreatment are high although substantial economies of scale exist, but the disposal costs for firms connected to the sewerage are also likely to rise substantially. These increased costs will make firms reconsider their water usage and there is already evidence of process substitution by dyers and finishers. Although costs may be comparatively small they may create substantial short run cash flow problems which the independent firms are less able to stand than the group members; group members also tend to use less water than independent firms and are able to draw upon group expertise in pollution control. Further, the move to equalise charges over all parts of the region will affect some firms more than others.

References

1. Harris and Garner, op.cit.

2. J.F. Garner, The Law of Sewers and Drains, Shaw, 1969.

3. The Solicitors' Journal, 13th June 1975.

4. Severn Trent Water Authority, Water Quality, 1974.

5. River Pollution Survey of England and Wales, 1970.

6. Annual Report on River Quality, 1974, Severn Trent Water Authority.

7. River Pollution Survey of England and Wales, 1970.

8. Annual Report on River Quality, 1974, Severn Trent Water Authority.

9. Private communication.

10. Described further in River Pollution Pt.III Control, L. Klein, Butterworths, 1966.

11. Private communication.

12. Private communication.

13. Broadsheet No. 9, BCIRA, Alvechurch, 1969.

Chapter VI

SOLID WASTES

The Legal Framework

Until 1972 the disposal of solid wastes was controlled only by the Litter Act 1958 and the Dangerous Litter Act 1971, but in that year, largely as a result of great publicity in the media, the Deposit of Poisonous Wastes Act passed through Parliament in the space of 22 days (1). This Act, and Orders made thereunder, lists types of poisonous waste and a general prohibition is imposed upon any deposit "liable to give rise to an environmental hazard". The Act also imposes a duty to keep records and requires local authorities to be notified of any such deposit (2). The Local Government Act of 1972 provided that the disposal of waste became the responsibility of each county authority in 1974 following local government reorganisation.

The DPWA therefore has two main facets, the imposition of penalties and the notifications procedure, but there was "no wish generally to stop the tipping of wastes on land. In particular, industry must be able to dispose of its waste and, in doing so, should not be forced to adopt complex practices which may not be necessary and which might be prohibitively expensive." (3) Thus there are a number of exemptions from the notifications procedure - listed in Statutory Instruments 1972 No.1017 - amongst which is the heading "sands including foundry and moulding sands" and although "it is to be expected that local authorities ... will adopt a positive policy of ensuring that tips receiving toxic waste are operated in accordance with the best advice available rather than a negative policy of unnecessary restrictions" (4), as we shall see later, these exemptions do not seem to have been universally applied.

The Control of Pollution Act 1974 seeks, in sections 1-30, to modify the law on solid waste disposal but, as at January 1st 1975, these sections had not yet come into force (5) and since most of our interviews were conducted before this date it is not relevant to set out the way in which the law may now have changed for the firms we interviewed.

Replies from Firms

The above considerations apply, of course, to the disposal of wastes by firms; we were also concerned about the types of wastes firms produced. Table 6.1 shows the types of waste reported to us by firms in the random sample; naturally some firms produce several different types of solid waste whilst three firms we interviewed produced none at all (one of these was a firm processing waste rags sent to it by other firms). To a certain extent the nature of these waste products is a reflection of the industrial classifications of the firms we visited but the one thing that stands out most clearly is that paper and packaging is reported by a very wide range of manufacturers. Not just printers but dyers and finishers, iron founders, engineering companies and food manufacturers all reported that paper or packagings formed some part of their waste. We were unable to obtain detailed

TABLE 6.1 Types of Solid Waste Produced by Firms

	Number of Cases Reported
Wood and timber	5
Scrap metal	14
Sand	9
Paper and packaging	20
Fabric	16
Leather, rubber or plastic	8
Sludges and chemicals	6
Oil	1
Other (including dust and rubble)	12
	91

information from firms on the amounts of waste of each type produced but it was clear that in the smallest firms amounts were, indeed, very small but in the largest firms waste management imposed distinct managerial requirements on the firm. In the case of one iron founder, a fleet of lorries was required to tip several hundred tons of waste material every week. Although a comparison of the different weights of materials might have been useful, the problem caused by solid wastes is not a factor of weight alone since, for example, because packaging tends to be very light for its volume, it may cause a bigger logistical problem to the space-starved firm than other, denser, wastes. This was the case in one firm of dyers and finishers which, whilst acutely short of space, had to accumulate large numbers of polythene containers before a contractor would contemplate removing them.

We were concerned to learn about the extent to which potential recycling possibilities were being exploited by industry. Just about one half (thirty-three) of the firms said that some of their waste was recycled or used as an input by some other manufacturer; of the remainder, fifteen firms said that none of their wastes were recycled and the rest said that they did not know. We were somewhat surprised at this apparent ignorance of businessmen about what happens to their wastes and, indeed, it would have been higher had it not been for the fact that we have included in the positive response to recycling cases where it was obvious that removal of waste by a contractor led to recycling such as in the case of metal recovery. Of course, there may have been more such cases that we have missed but we believe them to be few. Most instances of recycling occurred with metals and textile materials but we also found cases where sawdust was sold to local farmers and three cases where food wastes were also sold to farmers. In very few instances were wastes recycled by the firms that produce them. Examples of cases where firms actually did this were one engineering company which had found the local authority waste collection service so bad that it had been stimulated into recovering all its own waste and a firm making plastic soles and heels for shoes which installed a granulator to chop up waste PVC material where previously it had used an outside contractor to carry out this service *. We must stress, however, that these two cases were the exception rather than the rule and the degree of internal or vertically-integrated recycling appears very small. We can make several suggestions why this is so :

* This machine was so cost-effective that our respondent calculated that it amortised itself over a period of three months.

i. the ease with which waste products may be disposed;
ii. firms do not have the necessary technical expertise needed for recycling; and
iii. there are large economies of scale in reprocessing wastes and small firms in particular do not have a throughput large enough to justify expenditure on plant and equipment for reprocessing.

Naturally (ii) and (iii) are bound together; let us take as an example a small printing firm. It is obvious that paper making is a highly skilled and capital intensive activity which is clearly beyond the scope of a printing firm; the degree of skill - and possibly also capital intensity - is probably considerably enhanced when heterogeneous waste products are being used as raw materials, further reinforcing the inability of the small firm to process its own waste. Similarly with firms in the engineering sector; many do not possess the specialist skills or melting furnaces necessary to process their own scrap metal and we have found that even quite large secondary metal manufacturers send their drosses to other secondary producers because, amongst other reasons, they alone possess the necessary equipment for metal recovery. We include point (i) - the ease with which waste products may be disposed - since we feel that waste disposal is a peripheral activity which many firms carry out without great attention being paid to it and waste recovery through recycling often only comes about through some outside stimulus such as a substantial change in the prices of raw materials or because of increased difficulty (and cost) of disposing of wastes. Since the 1974 Control of Pollution Act and the 1972 Deposit of Poisonous Wastes Act, tipping space has become more and more limited and, as reported in The Sunday Times (6), contractors are finding it harder to find places to tip. Inevitably, this will raise the cost of waste disposal and in turn enhance the viability of waste recovery activities.

The reasons why firms did not recycle at all can be seen in similar terms and we must also add a further factor - the existence (or lack of it) of a suitable market for the wastes. The costs to the firm of separating wastes prior to recycling may exceed the costs of waste disposal (which may be zero for very small firms); firms may not know how their wastes can be treated or firms are unaware of the potential benefits of recycling. We found instances of cases where firms fell into one or more of these categories, instances where our impression was that separating waste material prior to disposal was "too much trouble" or where managers were quite unaware of how much could be obtained for waste materials. In other instances, however, the dramatic rises in raw material prices provided a stimulus to the awareness of some managers on the potential for waste recovery; for example, one firm which had previously simply thrown away paper and plastic sacks was seriously contemplating the employment of a person to segregate paper from plastic, thereby obtaining a high value for the plastic waste. Another firm began to recycle all its waste products when the local authority collection service became so bad that disposal was presenting the firm with acute logistical problems. Some firms experience no difficulty in establishing a market for their waste; we understand that agreements between metal suppliers and their customers that the suppliers buy back swarf and other waste metal are fairly common. Naturally such agreements are beneficial to both parties but are sometimes initiated by the customer. In one firm a change of management had brought with it a change in the stock of information on recycling opportunities; previously disposal costs had been in the region of £2000 per annum but, with the advent of recycling within the firm, this had been transformed into an annual profit of £3000.

Thus we consider that recycling is not generally an activity that firms automatically include as part of the production process - some outside stimulus such as pollution control legislation or a rise in prices is needed to promote the activity. Few, if any, waste recovery operations are self-financing and only changes in the relative costs of raw materials or in the cost of alternative means of waste disposal can bring such operations below the breakeven cut-off firms employ. Further, recycling is a function of the amount of information managers have available; at times of rapid inflation there is a tendency for information about prices to flow more rapidly than at other times. Thus inflation can help the information flow but so can government and industrial organisations and we therefore welcome the government's establishment of the Harwell Waste Exchange scheme which can only further increase the information flow to industry about the amounts and nature of wastes available.

Table 6.2 shows the number of instances of the four waste disposal methods reported to us.

TABLE 6.2 Waste Disposal Methods Employed by Firms

	Number	Percentage
Burning	8	10
Dumping	12	15
Local authority	20	25
Private contractor	40	50
	80	100

NB - Number sums to more than 62 due to multiple responses.

As can be seen, by far the most common method was the use of a private contractor. We were surprised that so many as twenty firms used the local authority but it must be borne in mind that only exceptionally small firms relied entirely on the local authority for their waste collection services; in most cases the types of waste reported under this heading were canteen wastes and small amounts of paper or packaging materials. Although only twenty firms told us that they used the local authority waste disposal services, we also asked firms that used them how they rated them and, in fact, thirty firms replied to this. The divergence in response rates for the two questions may be explained if firms do not regard, say, canteen or office refuse as part of their production wastes. Whilst twelve firms reported that they dumped wastes, this figure underestimates the degree of tipping of industrial waste since, although in some instances private contractors were taking wastes for recycling, in the large majority of cases they were acting merely as transporters of the waste between factory and tip. The least common form of disposal was burning and we found that it was only the smallest companies that used this and timber, sawdust or paper were burnt. No firm admitted that it burnt "nasty" wastes such as plastics or, say, rubber tyres. To an extent we were surprised that the prevalence of burning was as high as it was; we feel that burning (at least on the premises rather than at a local authority incinerator) represents a "disposal of last resort" which must indicate the unsatisfactory nature of local authority waste collection services.

We asked firms using the local authority services how they rated them; the results are shown in Table 6.3. Of course, these data relate only to those firms using the local authority services; one firm told us that it had found them so bad that they had been abandoned and others may well have had similar experiences. Others may have been too big altogether - as we will show later it is possible for the output of even quite modest firms to swamp local authority facilities. Nevertheless, the data still do not represent

TABLE 6.3 Rating of Local Authority Services

Rating	Number of Respondents
Bad	6
Poor	4
Average	10
Good	10
Excellent	-
Total	30

unqualified approval for the services offered to industry - twenty percent of those that use them rate them "bad". We heard numerous complaints from businessmen about local authority services; amongst them were :

i. that no industrial "skip" service was available such as that provided by private contractors or that insufficient bins were available,
ii. that the services were unreliable or that no "on demand" service was available that could cope with peak problems, and
iii. dustmen were untidy in their work or demanded bribes for collecting more than one bin.

Naturally we were not in a position to be able to substantiate any of these claims but we were left in no doubt that there is considerable potential demand for public services that emulate those of private contractors. However, before wholeheartedly endorsing suggestions for such a service, we would make the following points :

(a) We again refer to what we have already said about the way in which the firm reacts to the information flow it receives. Some local authorities told us that they do provide a skip service and yet firms were not aware of this.

(b) The volume of output of solid wastes of some firms is so great that existing local authority resources may be swamped by a requirement that the authorities take on this type of work.

(c) Local authorities might give consideration to improving their collection facilities for paper and packaging or other combustible refuse. This might lead to a higher recycling of such materials in the first place or more efficient burning in incinerators or district heating schemes in the second. This would arise since the industrial waste can be expected to be in large volumes and much more homogeneous than domestic refuse.

As has been seen, however, the most common disposal method was to call in an outside contractor; we were anxious to evaluate the costs of disposing of wastes in this way and what happens to such waste. From the point of view of the contractor these are, partially at least, two aspects of the same problem since if he finds the wastes easy to dispose of, he is more likely to charge a low price than if he finds difficulty in disposing of them. Materials that are easy to dispose of are ones that can either be recycled (such as metals or fabrics), burned (such as paper or plastics) or safely tipped. Many of our respondents told us that their wastes went as far as Pitsea in Essex and the reason is that one of the very few safe sites for tipping toxic wastes is located there. The one closest to the East Midlands is situated near Birmingham but none of the firms we contacted told us that they used it. Although some of the waste disposal contractors are installing sophisticated equipment for dealing with toxic wastes, the installed plant capacity is very low and disposal costs are high (7) - this may encourage firms to use the services of illegal fly tippers and whilst we had no evidence that any of the firms in our sample were using them many respondents professed ignorance as to what the private contractors did with the waste they collected. Three firms had been required to obtain tipping permits under the DPWA and there may well have been more firms that were tipping toxic wastes in small volumes in ignorance.

In addition to the actual disposal cost, firms will have to pay transport costs and skip rental. Naturally there is a very substantial fixed cost element in these charges and we feel that such fixed costs are a major deterrent to the very small firm considering using private contractors to dispose of wastes.

We asked firms to give us some idea of the costs of disposal of solid wastes and, if possible, the value of any waste materials sold by them. Forty-two firms were able to give us data on this and, of these, four firms actually made a profit averaging £1500 per annum. In all of these cases, the wastes being disposed of were metallic; for example, one was a firm using very large amounts of solder in its production processes. The solder droppings were collected by an industrial sweeper and returned to the supplier for reprocessing. Whilst the firm was doing this partly out of commercial reasons, it must be admitted that it also carried out the sweeping operation for health and safety reasons since solder contains large amounts of lead.

Whilst many of the other firms gained some benefits from the sale of scrap materials (most commonly metal), some respondents were not sufficiently informed to be able to tell us how much this amounted to. This fact reinforces our contention that waste disposal - like other activities in the firm - is a managerial function that requires a substantial flow of information about opportunities (both physical and financial) for recycling. In six of these 38 firms we were told that the disposal of solid waste cost nothing and the sums indicated by the others ranged from £8 per annum (for extra dustbins) to £10,000 per annum for a ferrous foundry firm. The average disposal costs of these firms was little more than £900 per annum - not a very large sum at all.

The above paragraph would suggest that disposal costs for ferrous founders may be much higher than for other firms and when we looked at our selected sample of iron founders we did indeed find that this was so. Iron founders have a solid waste disposal problem connected not directly with raw material inputs or with spoilt items specifically but with the sand used to form the moulds in which the molten iron is cast. In the larger plants very large volumes can be involved (several hundred tons per week) and, although it is possible to reuse some of the sand (especially if a sand reclamation plant is installed), ultimately the waste sand can be disposed of only by dumping since it has no other use.

The dumping activity has caused difficulties to some firms following the enactment of the DPWA; some local authorities, anxious to avoid the tipping of any toxic wastes have prohibited firms from tipping resin-bonded sands at various sites. The firms involved told us that such sands were quite inert and that there could be no rational justification for these moves and that some had been forced to continue to use the less efficient green sand instead as a result. Only three out of the twenty-four iron founders who discussed their solid waste disposal problems with us had any form of sand reclamation plant and all of these were in the larger size categories; this would suggest that there are substantial economies of scale in sand reclamation. Two of these firms also used their own tip for dumping although one said that it expected its tip to last no more than five years and when that happened it would experience considerable difficulties in finding an alternative site. All the other firms used the local authority tip and all but one had to pay to use it; that one exception was the only firm that ever reported to us that it had an "excellent" relationship with the local authority.

These solid waste disposal problems of iron founders are of quite a different order of magnitude to the problems encountered by other trades. This can be seen most clearly when looking at disposal costs; seven out of the twenty-four iron founders reported disposal costs in excess of five figures and the average cost of solid waste disposal for iron founders was £9500 per annum - more than ten times that for the firms in our random sample. It is now technically possible to reprocess some types of moulding sand using a sand reclamation plant but we had no evidence that the operation of such a plant would actually reduce disposal costs. As we have already said, only three founders had such plant and, although we were not able to obtain detailed cost information from all of them, our understanding is that both operating and capital costs are high even though these can be partially offset by reduced raw material and disposal costs. There do appear to be substantial economies of scale in this area which we feel small founders will not be able to reap. Further, we must emphasise that, whilst the idea of sand reclamation plant may appear highly desirable to environmentalists, casting sands are not in great scarcity nor does their disposal seem to present acute environmental problems. In addition, these plants bring with them the added problems of dust and noise and could therefore add to rather than subtract from the environmental impact of ferrous founders.

We also asked the textile dyers and finishers that we selected whether they had any solid waste problems; nearly ninety percent of them told us that they had some solid wastes. In most cases this was paper or polythene packaging and in a few cases other materials like textile waste, sludge and even aluminium swarf were reported to us. Of these, half knew that their wastes were recycled in some way but in all instances this was through a sub-contractor - none did it themselves. Most of the dyers used a private contractor; only one told us that it dumped and only five used the services of the local authority. Two firms thought the local authority service "bad" and the other three thought it "average"; this again represents a substantial vote of no confidence in local authority services since we gained the impression that many more would use them if they provided good enough facilities. The average disposal costs for dyers and finishers was approximately £1300 per annum - slightly higher than for the random sample, possibly due to the fact that the average size of firm was slightly higher amongst dyers and finishers than the random sample.

We were concerned to assess the extent to which firms had been affected by the Deposit of

Poisonous Wastes Act 1972. Out of our initial sample of sixty-two firms, only three said that they had been directly affected by the Act. There was, in addition, one ferrous founder which told us that it had been affected by the Act. In the main, the impacts on these four firms were of a very minor nature; one had been required to make a declaration to the waste removal contractor as to the precise nature of the waste and another had had to make a similar declaration to the local authority. The other firms had either had to change the place where trade waste was tipped (in the case of the ferrous founder) or install a settling tank to remove cutting oils (as in the case of a general engineering company).

It may seem surprising that the effects of the DPWA were not more widespread but, in fact, we could argue that some of the four cases mentioned above were not wholly within the scope of the legislation or were the result of panic or irrational thinking on the part of the local authorities. Given the nature of the activities of most of our sample of firms, we are not surprised that so few were directly affected; only a handful of the sixteen engineering firms were handling toxic chemicals and the impacted firms in ferrous foundry were using resin-bonded sand which, although novel, is probably quite harmless. Of course, it may be that some firms were using some chemicals in very small quantities and that their waste disposal contractors were unaware (unconsciously or consciously) of the nature of the wastes handled and it would be surprising indeed if any firm guilty of a contravention of the DPWA had told us about it.

Nevertheless, we have referred above to the direct effects on firms of the DPWA; we believe that the indirect effects have been very much more widespread. Local authorities have been so reluctant to grant permits for tipping sites that, as we have already explained, most industrial waste is transported very large distances to Essex. Whilst this may be all very well for those areas which are potential tipping sites, it is not very good for those authorities through which the waste passes and worst of all for Essex. Furthermore, it probably increases waste disposal costs for all firms regardless of the harmfulness of their wastes. Pollution is an emotive issue and it seems to us that this intransigence of local authorities - which adds to industrial costs - is a result of irrational and ill-informed decision making on their part.

Local Authorities

Naturally in our approach to local authorities we asked them their attitudes to the collection of industrial waste. Already, of course, the authorities do collect light trade refuse which in most cases amounts to canteen waste but few operate any scheme for other trade refuse. Our discussion indicated that local authorities had begun to study the problem, in anticipation of increased demand for help by industry. We were referred to a study (8) which, whilst it largely confirms what we have said earlier in this chapter, reveals some interesting facts about trade waste.

The survey covered over six hundred firms in the Manchester/Salford area and their industrial breakdown is shown in Table 6.4. Whilst we may expect the Manchester area to be slightly different from the East Midlands - with more heavy industry in the former - structurally the regions should be comparable. Naturally the report does indicate that the precise nature of the industrial wastes differs at various locations according to the nature and size of industry. The total waste generated by industry in the Manchester/ Salford area was more than twice the weight of the domestic refuse handled by the local authorities so there can be no doubt that even if half the industrial waste were to be handled by local authorities, their current facilities would be swamped.

TABLE 6.4 Industrial Breakdown of Firms in Trade Waste Survey

Category	Proportion (%)
Food	5
Light chemical	6
Heavy chemical	5
Metals	6
Electrical engineering	6
Textiles	8
Clothing	20
Bricks (and building)	6
Timber	6
Paper	9
Mechanical and general engineering	22
Others	6

The authors of the report were concerned, of course, with the ways in which the wastes may be disposed; their findings were that most of the four fifths that was not combustible would need to be tipped and one fifth could be incinerated (paper, sweepings, plastic, timber and sawdust, etc.). Approximately one quarter of the waste was combustible or dangerous, being toxic, inflamable or caustic; the sludges needed special treatment prior to disposal.

Not surprisingly, the largest firms produced most wastes but it was also clear that the firms employing more than 500 people had a waste production rate (in terms of tons per employee per year) of approximately three times that of firms in the other size groups whose waste production rates did not differ significantly. We find this fact surprising; unless the data have been severely adversely affected by results from one or two firms, then this clearly indicates that either different types of product are made by large and small firms or that different processes are used by them. If the latter, then it could be that higher labour and capital productivity rates result in the same waste/unit output ratio being translated into a higher waste/employee ratio. However, such studies that have been made of labour and capital productivity have not revealed differences large enough to account for a difference of this size and we must therefore conclude that more waste intensive processes seem to be used in large firms than in small firms.

In the sample it was found that the three largest producers of waste were bricks, heavy chemicals and metals, in that order. If we take account of the fact that our survey was not intended to cover construction (which was included under bricks) and that, since we were looking only at small firms, we did not locate any heavy chemical firms, this report confirms our findings that metal manufacture is the heaviest producer of waste amongst small firms. The output rate for metals was approximately three times that for firms in the food, light chemical and paper and printing trades which were, in turn, about three times that of firms in the electrical, textiles, clothing, timber, engineering and other trades.

The disposal methods utilised by firms differed little from those we found and are summarised in Table 6.5.

TABLE 6.5 Proportion of Different Methods of Dealing with Waste

Tipping		91%
of which,	removed by contractors	60%
	tipped on own site	30%
	removed by local authority	10%
Incinerating		2%
Burying or dumping at sea		7%

These data accord well with the results that we found, although there was a higher incidence of local authority assistance; 51% of firms said that they received some local authority assistance but the report goes on to say that the majority of firms had less than 100 employees and ".... it is safe to assume, therefore, that most of the refuse concerned could more accurately be described as trade waste, which would in any case be removed on normal bin rounds" - we feel that on the basis of our experience this is somewhat optimistic. The incidence of incineration is also interesting; one fifth of firms carried out some incineration but this generally amounted to less than 10 tons per year. The data published in the report on other methods of disposal are obscure; we take them to mean that some of the waste removed by private contractors is also dumped at sea in sealed containers in addition to the 7% of waste that is disposed of directly by firms in this manner.

Since the Manchester/Salford study was completed, other studies have been conducted elsewhere (for example Wiltshire and Hampshire) which have broadly confirmed these results except for the fact that the precise nature of industrial waste depends upon the nature of the industrial mix in the area. The Wiltshire survey also provides information on the amount of recycling (or "salvage" as the report calls it). The greatest weight of salvaged material was in ferrous metal where six and three-quarter thousand tons per annum were salvaged out of a total weight of eight thousand tons, giving a salvage rate of 84%. Roughly two thousand five hundred tons of both foodstuffs and paper were salvaged, giving recovery rates of 71% and 26% respectively. The highest recovery rate appeared to be in non-ferrous metals where all the 750 tons or so produced annually were salvaged. The greatest scope for increased recycling seems to be in paper and card where not only is the rate of production very high, but also the combined proportions of tipped and incinerated materials represent 71% of the total.

Whilst local authorities have been concerned about trade refuse, it remains true that few have done anything about it, probably because they realise that demand is likely to outstrip the supply of services. But, of course, the local authorities compete with private contractors for tipping space and a continued policy of non-intervention may be self-defeating. Even if improved waste recovery were not possible, the use of controlled incineration for district heating is possible and being carried out in some parts of the country (particularly Nottingham). It may be that there would be scope for the increased use of combustible trade waste as a cheap source of fuel for such incineration schemes.

Conclusions

Although the precise nature of solid wastes in any region depends upon the specific nature of the region's industrial mix, most firms create waste paper or packagings which are often difficult to dispose of. Many firms are taking opportunities to recycle their wastes and, although more could possibly be done, this is almost universally carried out by third parties. Very few firms recycle their own wastes due to the economies of scale and specialised nature of the waste recovery processes. The extent of recycling depends upon the awareness of firms about the technical and financial opportunities; ultimately this depends upon the standard of management and the way the firm organises the flow of information within it. Inflation can act as a stimulus to increase recycling. Most firms resort to private contractors for their waste disposal and do not rate the local authorities highly although there exists a substantial demand for their services.

References

1. For a record of the circumstances leading to this Bill see Kniber, Richardson and Brookes, Deposit of Poisonous Wastes Act 1972 : Government by Reaction?, *Public Law*, 1974.

2. Summarised from D.A. Bigham, *The Law and Administration relating to Protection of the Environment*, Oyez Publications, 1973.

3. *Circular No. 70/72*, Department of the Environment, 19th July 1972.

4. *Ibid*.

5. See *Lumley's Public Health*, Vol.XVI, 1974.

6. *The Sunday Times*, 16th May, 1976.

7. *Ibid*.

8. *Getting the Measure of Industrial Waste*, Local Government Operational Research Unit, Royal Institute of Public Administration, April 1973.

Chapter VII

NOISE

The Legal Framework

Noise was defined by the Committee on the Problem of Noise (1) as "sound which is undesired by the recipient". For present purposes, we can distinguish two classes of recipients of industrial noise - employees of firms and third parties (probably nearby residents). For control of noise levels inside the plant as they affect workers, the major control seems to have been, until 1974 at least, that exercised by the Factory Inspectors through the Factories Act 1961. However, with the advent of the Health and Safety at Work Act 1974, it seems that the control of noise within factories will be tightened by the executive with its ability to impose "improvements" and/or "prohibition" notices.

For third parties affected by noise, redress may be found by a common law action in nuisance provided the noise is sufficient to cause "substantial interference with health, comfort or convenience". The Noise Abatement Act 1960 declared noise to be a statutory nuisance under the Public Health Act of 1936 and thus enabled local authorities to take the initiative in noise control without the need for prior complaints from the public. However, in deciding such a case, the courts would take into account whether the "best practicable means" had been employed to abate the noise and, if so, the cost of further means necessary to remedy the trouble. Control of industrial premises was also exercised at the planning stage by a variety of planning acts such as the Town and Country Planning Acts and the Public Health (Recurring Nuisances) Act 1969. However, the Control of Pollution Act 1974 seeks, in sections 58 to 74 inclusive, to modify and extend much of this legislation. For example, it replaces the Noise Abatement Act 1960, which is to be repealed. Further, section 68(1) (which had not been brought into force by the end of 1974) specifically deals with noise from plant or machinery and permits the Secretary of State to lay down standards and specifications for machinery, but section 68(2) also provides that the Secretary of State shall ensure that :

> ".... the regulations do not contain requirements which, in his opinion, would be impracticable or involve unreasonable expense. It shall be a defence if 'best practicable means' have been employed for control."

Replies from Firms

When we asked firms whether they had any noise problems, we were acutely aware of the fact that we were not experts in this field and we had no intention of going into firms and measuring noise intensity. The analysis of noise is, of course, a highly specialist area and exhaustive noise measurements were quite beyond the scope of our study and we therefore have had to rely on a quite different approach which is, in fact, quite consistent with the objectives of the whole study. We can define noise as "unwanted sound" and, if we do this, it becomes apparent that, depending upon the circumstances and preferences of the individual, many sounds may be termed "noise" which in terms of some objective criteria may not be. The sound of a typewriter in an adjacent office may be perceived by some as noise since it disturbs concentration but, on the other hand, some industrial trades may not perceive noise problems (even though sound levels may exceed objective criteria) simply because the noise has come to be "accepted" as part of the trade (in economic terms, the demand function has shifted outwards). In asking firms about noise problems in their factories, we were therefore asking them about their perception of noise. This does not mean that we can necessarily accept negative replies to indicate the absence of a noise problem since there is much evidence that noise may go unnoticed even though its effects are serious. For example, Jean Stone writes in a study of hearing of hosiery workers that long-term exposure to noise can cause industrial deafness and the insidious nature of this condition is that it is at an advanced stage when the hearing of speech becomes impaired (2). Thus it may be that firms do not perceive a noise problem in the workshop (or elsewhere) even though the noises are of a sufficient intensity to cause hearing loss.

Of course, much of the above relates to the internal working environment of the firm; firms can, and some do, have noise problems associated with the outside environment, particularly residents in nearby houses. We have already shown how the legislation copes with this situation but let us also add that we believe that in some circumstances firms might perceive external problems where internal problems are not perceived. This could happen if, for example, workers accustomed to noise in the workshop were content but nearby residents did complain of, say, nighttime working or where air-conditioning plant inside the factory could be heard outside.

In our study of noise problems, we were therefore concerned with both internal and external noise and in the Tables which follow we deal with all firms together. Table 7.1 shows the incidence of internal noise problems as perceived by industry :

TABLE 7.1 Incidence of Internal Noise Problems : Proportion of Firms in Each Trade Perceiving Internal Noise Problems

Type of Trade	Number	Percent
Textile dyeing and finishing, clothing and footwear	17	40
Engineering	9	56
Metal manufacture	24	65
Plastics	0	-
Timber	2	40
Food	1	25
Building materials	0	-
Printing and paper	1	-
Other	3	60
Overall	57	48

As can be seen, the highest incidence rates are in engineering and metal manufacture - as may be expected. Overall, about half of the firms interviewed had some sort of internal noise problem. Many of the noise sources were not unexpected - drilling, grinding, planing or sawing equipment for wood or metal, generators and compressors, steam injectors and, in foundries, shake-out and fettling machinery. The latter relates to the ferrous foundries where a BCIRA report (3) shows that some areas of the foundry (dressing, moulding and shake-out) can have sound levels close to, if not exceeding, that of an unsilenced road drill at 7 metres (90 dBA).

Another item of equipment referred to us was the knitting machine and, in this area, the report of Jean Stone (op.cit.) also gives some valuable evidence. The report, which covers two factories, states that "some of the highest noise levels within the industry are associated with the process of flatbed knitting and circular knitting" and it therefore looked specifically at these machines. In the first factory 11 out of 16 machines exceeded a noise level of 85 dBA and six exceeded a level of 90 dBA (the equivalent of the road drill mentioned earlier). In the second factory only one machine had a noise level less than 85 dBA; all machines had noise levels exceeding 80 dBA - equivalent, say, to an alarm clock at one metre. Mrs. Stone continues that many of the existing damage risk criteria would rate noise levels exceeding 85 dBA as being hazardous and, in these two factories, therefore, nearly 80 percent of the machines surveyed would be deemed hazardous.

There were also cases of more unusual or situation specific noise sources; we have addressed ourselves to one firm in detail in Case Study 4 in Chapter VIII where the firm was running motor engines for extensive periods both day and night. For technical reasons it was impossible to soundproof the engines and therefore extensive measures had to be taken to protect employees and the outside environment.

Another firm reported that it had acute noise problems from seeds falling on to metal sheeting after they had been coated with chemicals, and a firm of bell manufacturers told us that difficulties were encountered in the bell-tuning process. Another frequently reported noise source was the equipment used for improving the air quality in the workshop; it seems that the installation of fans to extract air quite often leads to an undesirable noise level. This is clearly the sort of situation where firms trade off one sort of pollution for another; an odour problem that cannot be dealt with by gas cleaning may be attenuated by ensuring that the fumes are discharged at a high level. We have encountered one firm which occasionally experienced odour problems and claims that the only way to deal with the problem is by high discharge, which would require large noisy fans which would, in fact, be less acceptable than the odours. We came across one further instance where pollution control equipment had had an unfortunate noise trade-off; the granulator bought by the plastics manufacturer which allowed him to recycle his own PVC material was extremely noisy and special provision had had to be made for it.

Table 7.2 summarises the way in which these internal noise problems had been tackled; in half the firms the solution used had been that of issuing ear muffs. In some circumstances this is satisfactory since there are some pieces of equipment that cannot be silenced (as in the case of the firm in Case Study 4), and it does provide the firm with a very low cost control method, providing that workers can be persuaded to wear them. The firm in Case Study 4 had paid particular attention to this aspect of the problem and had devoted much management effort to schemes to get workers to wear their ear muffs. This firm had acute problems calling for radical solutions; we gained the impression that some firms with less acute problems were not so willing to put so much managerial effort into solving noise

TABLE 7.2 Measures taken to Attenuate Internal Noise Problems

	Percent
Ear muffs	50
Insulation	31
Change in process	4
Nothing	15
	100

problems. Once again, this is a result of differences in perception of the noise problem which is a key parameter in determining the firm's response to pollution. Whilst many firms told us that ear muffs were "freely available" especially if "asked for" by the workers, we found few other instances of campaigns to make staff more aware of the dangers of noise or to persuade workers to continue to wear their ear muffs. Indeed, in some cases there was antipathy to ear muffs - mainly under the misguided understanding that they could be positively hazardous if workers could not hear shouted warnings of danger.

One firm told us that it was concerned about the possibility of prosecution under the Health and Safety at Work Act if, despite all measures to encourage workers to wear ear muffs, they continued not to do so and subsequently sued the firm for damages arising through industrial deafness. We discussed such issues with representatives of several trades unions; although there appeared to be considerable apathy, the Textile and Hosiery workers had recently commissioned a study of noise problems (that by Mrs. Stone) and the General and Municipal Workers Union had recently won the first noise award damages in the UK. On the whole, however, these were isolated instances and unions concentrated their activities in the area of wage negotiations.

Another way of coping with noise in factories is to isolate the source and, perhaps, move the noisy machinery to separate parts of the building or to soundproofed areas, thus minimising the number of people exposed. As Table 7.2 shows, 31 percent of the firms had used some such sort of acoustic insulation to shield workers; in some cases this had been accomplished at low cost where, for example, it had been a matter of removing a compressor or a generator to a separate building. In other cases the noise problem can be solved by some change in process - different equipment or raw materials, say. Three firms had done this; in one, machinery had been modified by substituting fibre gears for metal ones and in another PERA had advised the use of another type of metal in the production process.

Twelve firms in all told us that they had experienced problems with external noise pollution; again, the most important of those was Case Study 4 which had designed and built its buildings to stringent noise criteria. In addition to this firm, there were three in textile dyeing and finishing, six iron foundries and two others; these all tended to be larger firms where the greatest difficulties arose with the nightshift. In some cases extractor fans were again to blame but in others - particularly the iron foundries - it was the production process itself. One firm reported that the piped music to its workshop had been the subject of daytime complaints from local residents. These cases of external noise pollution were harder to control than internal problems - the palliative of ear muffs is no longer available. Most of these firms had therefore employed noise consultants to help tackle their problems and had introduced more, indirect, noise control measures than simple acoustic insulation; for example, new plant was being designed to meet noise specifications, buildings had been specifically designed for noise control measures and in one firm the choice of machinery had been influenced by noise control criteria. Not all of these firms had actually managed to solve their noise problems and, for others, control costs had been very low, but an average of £8000 in terms of direct capital equipment would not be unrepresentative. One firm installing new equipment told us that the noise pollution control represented 15 percent of its equipment cost. In addition to these tangible costs must be added the intangibles of process changes, etc., which cannot be costed out.

The direct costs of those firms supplying ear defenders only was approximately £250 per annum - quite a small sum; in the iron founders noise control had cost an average of £5000 per plant in terms of capital equipment and running expenses were very small. Nevertheless costs were expected to increase substantially in many firms where more stringent controls were anticipated in the future (see Chapter IX).

We were also interested to know what impact, if any, noise control equipment had had on efficiency, particularly labour efficiency. Twenty-seven firms told us that it had made no impact, nine said it had reduced efficiency (mainly through communication with workers wearing ear muffs being impaired) and five actually reported that efficiency had increased - mainly as a result of reduced labour absenteeism and turnover. Naturally this is extremely subjective because we did not define what we meant by "efficiency" and we could not be sure that the replies we received were based on objective measurements by management. We do find it surprising that quite so many firms told us that they had found it more difficult to communicate with workers wearing ear muffs since we believe that modern devices do not attenuate sounds within the range of human hearing. We asked many questions of the firm in Case Study 4 but our contact said that their pattern of labour turnover was that staff either left very quickly or that they stayed a long time. Those leaving quickly gave a variety of reasons but noise did not figure greatly (other factors being pay, inaccessability, etc.) and amongst those who stayed a long time noise had never been quoted as a reason for leaving. Of course, this was an isolated firm but the role noise plays in the determination of the pattern of labour turnover obviously depends on the type of workers employed and their inherent mobility - skilled electrical fitters may be able to work almost anywhere but a skilled dyehouse operative may only be able to find work elsewhere in the textile industry which is contracting anyway. Thus we can come to no firm conclusions about the importance of noise in determining productivity or labour turnover.

Clearly noise is recognised as a substantial problem in a good number of larger firms; five had commissioned or completed noise surveys on their works by expert noise consultants. These surveys obviously play an important part in the firm's assessment of future costs; we have already seen how 80 percent of the machines in two textile knitting factories were probably hazardous and, if this pattern were repeated elsewhere, it would go a long way to explaining why many firms anticipated that their major pollution expenditures would in future lie in noise control. Communicating the dangers of noise to managers of smaller firms seems much harder, however. Some of the "intangible" costs of control could be very high; for instance the firm of bell-makers estimated that to move its foundry to an area where bell tuning would not affect residents would cost upwards of a quarter of a million pounds. Other firms did not perceive quite such difficulties - one recognised that soon the ear muffs it supplied would be insufficient and more would be needed to be done but others had different views. One respondent, justifying his lack of action over noise control, explained that his workers were skilled men who made few complaints because they were "basically craftsmen and there was good logic behind their thinking".

Local Authorities

As we explained previously, the Noise Abatement Act deemed noise a statutory nuisance, thus empowering the local authorities to act upon their own initiative. Our discussion with the local authorities showed that few act in this way, preferring to rely upon complaints from the public (as was also the case for odour and airborne emissions). Further, the authorities' willingness to use legal sanctions is even more limited; none of the authorities we visited had actually prosecuted any firm for noise reasons although at least two told us that they had issued abatement notices.

We can only speculate as to the reasons behind the authorities' reluctance to use legal sanctions; firstly, of course, it may be that the authorities do not have sufficient resources to devote to what they may well consider a low priority area and, secondly, it may be that the authorities wish to retain good relations with local industry and that such relations might be soured by their own initiation of investigations. The authorities gave us the impression that they prefer to use persuasion rather than compulsion and to act as advisers rather than policemen. Certainly many of the officials we spoke to in the authorities were well aware of the employment implications of some of the alternative noise control schemes mentioned and it must also be borne in mind that the councils in many instances provide an interface between industry and the authority when people from local businesses are elected to the council. However, in the council we thought most active in the field of pollution control there was no evidence to suggest that this was in any way a function of an "anti-industry" lobby.

The resources local authorities have available for noise control appear to be limited. We contacted the local authorities after the local government reorganisation (which may have been expected to have created more opportunities for specialisation of staff) yet none had a fulltime noise officer. Almost all the authorities had some sort of noise monitoring equipment (average cost £350) but our impression was that the equipment was not fully utilised because of a lack of suitably qualified personnel. Nevertheless, the smaller authorities with less sophisticated equipment were able to draw on a pool of greater expertise from the county councils and even the larger authorities were prepared to call in specialist noise consultants in cases where they lacked sufficient expertise. In this way, the absence of trained specialist staff had not hampered the investigation of complicated problem cases in the past; it may well, however, have hampered the investigation of more routine cases where the bringing in of consultants could not have been justified. There can be no doubt that there has been a technological revolution in noise and noise control in the recent past and now, more than ever before, trained specialist skills seem to be needed in this area. The evidence that we have from the local authorities confirms this; many of those in industrialised areas were sending staff on training courses at Leeds or Salford Universities or had established informal links with local colleges which were able to offer expertise. Thus although the authorities have been lacking in necessary expertise in the past, we perceive that they are actively and rapidly increasing the expertise in this area, but it will naturally remain a very small part of their total operations.

At present noise accounts for only a small proportion of complaints to the authorities; for example, in Hinckley where we were given a detailed breakdown of the complaints received in the first eleven months of 1974, only two percent related to noise. This is also true of Nottingham City Council; in 1968 only 0.8% and in 1972 only 0.7% of complaints arose from noise but if only the three categories of "pollution complaints" which could arise from industry are taken (odours, smoke, etc., and noise), then noise complaints account for about one quarter of the total. However, most of the authorities we spoke to told us that the number of complaints was increasing and they expected them to continue to increase. Respondents ascribed this to increased awareness of noise amongst the public and also partly to increased home-ownership, but one respondent told us that, as with air pollution, the fact that there were more complaints did not necessarily mean that noise pollution problems were deteriorating but could simply indicate a growing willingness on the part of the public to complain. Further it is apparent that some old established attitudes die hard; Derby District Council told us that they never got any complaints about collieries which, of course, have long been major employers in the area.

There was some slight evidence from our study that noise was more "acceptable" in the more industrialised areas and proved a more difficult thing to control in the less industrialised areas. For example, in one market town near Nottingham we were told of a case where the major local employer was engaged in a very large noise abatement programme involving large amounts of acoustic curtaining and thousands of pounds worth of trees to form additional sound screening around the plant. The only other authorities quoting individual cases were less industrialised ones like Spalding and Newark.

Of course, measures such as these may really be described as "end of line" pollution control and we found that many authorities were also actively engaged in stopping noise complaints at source at the planning stage. This control can have two facets; it can seek to reduce the amount of noise emitted by factories by requiring modifications before the plant is built - several councils told us that they were actively engaged in this sort of control through planning. Alternatively, it can control the number of people exposed to the noise by, for example, prohibiting residential development close to industrial sites and vice versa. Most authorities in the industrialised areas we visited were setting noise limits on factories being planned or were specifying the number of hours which a factory could operate, thus preventing nightshift working which can lead to a great many complaints. Some of the larger authorities acknowledged that noise control was one of the factors leading to the establishment of industrial estates. The amount of resources devoted to this control through planning are impossible to establish. However, we gained the view that it represents a comparatively small amount of the time of a planning department and, superficially at least, appears to be an effective form of control. Naturally the factory that is not built cannot cause a noise problem but that may mean fewer jobs and a national loss of output so there can be no definite assessment whether such control is justified.

Conclusions

We conclude that internal noise problems, at least, are fairly widespread in industry; nearly one half of the firms in our sample had such problems. Naturally some trades are inherently more noisy than others. Many firms are able to overcome most types of noise by supplying workers with ear muffs; these are not expensive but in some cases firms had more intractable noise problems that could only be dealt with by more expensive methods.

Problems with external noise were mainly limited to the larger firms and were often associated with nighttime work or with items of pollution control equipment such as air-conditioning. Control of external noise is expensive, necessitating in most cases the advice of specialists in view of the complexity of the topic. Many firms told us their concern with noise problems was growing; there is an increasing tendency to specify noise limits on new machinery, for example. Many firms thought that noise would be their major future cost area as far as pollution was concerned.

The local authorities do not devote large resources to control of noise; they tend to rely on complaints and these tend to be few. Nevertheless, many have access to specialist advice which they are prepared to call upon when needed and most authorities are operating cost-effective controls on noise at the planning stage of new developments.

References

1. Cmnd 2056.

2. Jean Stone, Noise and the Hosiery Worker, The Church Gate Press, 1971.

3. BCIRA, Broadsheet 69, BCIRA, Alvechurch, 1973.

Chapter VIII

SPECIMEN CASE STUDIES

Since the principal objective of our study was to look at the control problems individual firms faced, it is appropriate to provide some examples of the case studies that were written immediately after the visits by the research team. Naturally not all the visits were equally rewarding in terms of the pollution problems encountered; we have therefore excluded case studies on those firms which, by nature of their trade or size, had few, if any, pollution problems. Rather we present cases which at the time of the research seemed to be of outstanding interest for one reason or another. Also those we have chosen present a variety of technical problems and a variety of sizes of firms; although it would be possible to present case studies of all the visits we made to, say, dyers and finishers, the repetitious nature of the problems would quickly become tedious for the reader and we have thus selected only a few firms from each problem area or trade.

Naturally the information we collected from firms was offered voluntarily on the basis that it would remain confidential. We have therefore had to avoid disclosing the names - and in some cases the detailed nature of the trade - of the firms.

Case Number	Brief Details - From the Random Sample
1	Typical general engineering company with no large scale problems.
2	An engineering company with abnormally large liquid waste discharges.
3	A company involved in metal manufacturing with several facets of pollution.
4	An engineering company with acute noise control problems.
5	A large scale foundry.
6	A factory in the food trade with liquid effluent discharges.
7	A food processing factory with odour problems.
8	A textile finishing firm with stenter fume problems.
9	A large scale textile finisher encompassing the whole range of pollution problems.
10	A smaller scale finisher with some novel control methods.
	Brief Details - From the Sample of Dyers and Finishers
11	A very large independent company dyeing solely on a commission basis.

12	A firm similar in size to 11 but part of a vertically integrated chain and involved in no commission dyeing at all. It has few effluent problems.
13	A firm in a similar situation to that in 12 with the exception that it has considerable effluent problems.
14	A very small independent firm that was contemplating closure, partly as a consequence of pollution control.
15	A smallish firm with stenter fume problems.

Brief Details - From the Sample of Iron Founders

16	A large founder with sophisticated pollution control equipment.
17	A medium sized independent jobbing foundry.
18	A small independent jobbing foundry likely to have substantial impact costs.
19	A foundry which changed its type of melting furnace, partly in response to pollution control.
20	A medium sized foundry which has installed wet arresters but for whom pollution problems remain.
21	A very large founder with little pollution control equipment.
22	A medium sized foundry where the installation of pollution control equipment was likely to lead to the closure of the redundant plant.

Case Study 1

This firm, with between 50 and 200 employees, manufactures small machine tools and milling cutters and is located in what is virtually a "greenfield" site in the north of the region. Initial contact was with the Company Secretary who declared that "90% of the questionaire" was irrelevant, but an interview was arranged with the Works Manager which revealed that there were several pollution problems.

Although the firm is not located in a smokeless zone, all heating was oil-fired and the respondent thought that this did not cause any air pollution since he was ignorant of the sulphur dioxide content of the emissions from such heating systems. The principal air pollution problem encountered by the firm, however, arose in the heat treatment section which operated about 40 hours per week. Carbon monoxide, sulphur oxides and nitrogen oxides were given off although the first was in very small volumes. Face masks were provided for the staff working in this section but the respondent considered that the atmospheric pollution of the area around the factory was slight since the factory was working only forty hours per week. In addition to the provision of face masks and breathing apparatus, which the respondent said reduced workers' efficiency (particularly in hot weather), dust extractors were installed in parts of the factory (principally the fettling and paint shops). In all, approximately £10,000 of capital equipment had been installed with an annual expenditure of about £1500 (1973 prices) on air pollution controls.

There were two liquid pollution problems: the disposal of soluble cutting oils and the disposal of large volumes of dirty hot water from the cooling plant associated with the metal heat treatment plant. The soluble cutting oils were diluted 40:1 before use and the respondent stated that they contained no toxic materials; in fact, the local authority had agreed that these wastes could be discharged directly into the sewerage without treatment since there was no likelihood of toxic materials being carried in them. The waste hot water was recycled as far as possible but the worst contaminated water was discharged to the sewerage. Although this discharge was cost free, the respondent estimated that approximately £100 * of the annual water bill of £250 * was attributable to the cleaning of cooling water.

Three types of solid wastes were disposed of by the factory: Barium salts from the heat treatment process, steel swarf and general waste such as timber and other packaging materials. The Barium salts were disposed of by a private contractor; the respondent gave no indication of the charge made for this and had no idea of how the contractor disposed of what he collected. Since the passing of the Deposit of Poisonous Waste Act 1972, these Barium salts had been packed into sealed drums but, prior to this Act, this waste was disposed of by tipping. The steel swarf was divided into high and low cost types and was generally sold to a private contractor and subsequently recycled. The general waste was collected both by a private contractor and by the local authority; the firm paid the contractor £250 * per annum and, although the local authority collection service was free, the respondent thought their service could be improved in certain respects. For example, there was a reluctance to collect grinding waste, the punctuality of the service left something to be desired and it would have been more helpful if there were a skip disposal system.

Ear muffs were provided free of charge to all who requested them, but we were told that they were worn by only a few employees.

The respondent thought that pollution and noise control measures accounted for less than 5 percent of production costs but these extra costs were passed on to customers in the form of higher prices since they were considered to be part of overheads.

Case Study 2

This engineering firm manufactures automotive components, principally radiators and car heaters. The production processes involve chemical cleaning of metals and fluxing operations followed by oven baking or hot dip tinning; less than 200 people are employed in the one factory. Space heating in the factory is by oil and gas but the main air pollution problem arises from the fact that the cleaning processes give off hydrochloric acid fumes which are vented directly into the atmosphere by means of extractor fans; the firm believes that, nevertheless, all aspects of the Clean Air Acts are satisfied (it is not located in a smokeless zone). Further, large volumes of solder are used in the manufacturing process and since this contains large amounts of lead, the Factory Inspectorate requires that those persons who have close contact with the soldering processes have regular medical checks for the lead level in their blood. The firm did not indicate whether

---

* 1973 prices

checks on emissions outside the factory had been carried out.

Cleaning processes also use large volumes of water - currently approximately 16,000 gallons per day, which is discharged direct to the sewerage. This waste contains some acid (which is largely neutralised by caustic soda used at another stage), some oil and some metal residues. It is the latter which is the cause for some concern since it arises from the zinc chloride in the flux and the firm estimated that approximately 160 pounds per week of zinc are disposed into the sewerage without treatment and could, possibly, be recovered. At the time of the visit the local authority was disposing sewage on farm land but was building a treatment plant and the company feared that it would be liable for substantial sewerage charges once the plant was completed. The company was therefore actively seeking ways in which it could treat its own waste and recover the zinc being discharged; all contacts with contractors up to the time of the visit had suggested that such treatment plant would be very costly.

The production process gives rise to considerable volumes of metal scrap and solder dross; both these are recycled. The scrap metal (which is worth several thousand pounds per annum) is sold to merchants and the solder dross (which is recovered through vacuum sweeping machines on the workshop floor) is resold to the manufacturers. Some waste is removed by the local authority and, although the company rated the local authority service as "average", it felt that the service could be improved in terms of reliability and supervision and that the authority should have implemented the recovery of waste paper. The Deposit of Poisonous Wastes Act has affected the firm; prior to 1972 waste oil was buried in the ground but it is now collected by contractors (the costs of this service were not disclosed).

Although the workshop is noisy, only the generator room has been soundproofed and employees near to the sound source have been issued with ear muffs; this had had no effect on efficiency.

At the time of the interview the firm spent less than 5 percent of its production costs on pollution control measures but expected this figure to rise by between 200 and 500% in the future as a result of the opening of the sewage disposal works.

Case Study 3

The firm - a subsidiary of a large multi-national company - employs about 200 men and is one of the largest galvanising companies in the UK. The firm is not involved in metal manufacture but works on a commission basis galvanising materials brought to it by customers. The process involves four basic stages : (1) cleansing of the metal in a hydrochloric acid bath; (2) washing in cold water; (3) immersion in a flux solution of ammonium chloride; and (4) immersion in a zinc bath for coating and galvanising. The process is relatively simple, although fairly capital intensive. The split of total costs between overheads, raw materials and labour was roughly equal. About 1300 tons of zinc, 1000 tons of hydrochloric acid and smaller amounts of ammonium chloride were used annually. The heating and power used in the processes were derived entirely from gas and electricity. The process emits substantial volumes of toxic fumes (zinc ammonium chloride and ammonium chloride); the latter straight into the atmosphere from three 120 ft. high chimneys, but the former was filtered out by use of fume extraction equipment incorporating a ferrous catalyst. According to the Works Manager, the company was one of the few in the UK to incorporate this fume extraction equipment which cost £35,000; after its installation the men refused to work unless the equipment was operating. There had been no complaints from local authorities so far over the emission of ammonium chloride into the atmosphere, but the firm felt that control of such emissions would be politically difficult in the area. The other area in the factory where there was an air pollution problem was in the shotblasting shop where, at a cost of £2000, dust extraction equipment and face masks had been installed in 1967.

The discharge of liquid waste by the company took two forms; firstly the acid used in the cleansing process which was taken back by the supplier for recycling at a cost of £4.50 per ton - a total cost of £4500 per year. The other liquid discharge consisted of large volumes of water which was slightly acid and which was pumped straight into the sewerage at apparently zero cost to the company. The Trent River Authority had expressed some concern about this and the company was actively considering installing neutralising equipment since it felt that it would be forced to do so at some stage.

The solid waste took two forms. Firstly, dross (basically zinc/iron solids); this was sold to a private contractor and found its way back intc zinc metal by a resmelting process; the firm was actively considering its own recycling of this waste but felt that it would not be economic until the price of zinc had reached £400 per ton (it was £350 in 1974 and had risen above 250 percent in the previous two years). Secondly flux skimming; this was sold to a private contractor and often found its way into boot polish or cosmetics.

Pollution control played an important, though minor, role in cost terms. Pollution control equipment was seen as a fixed cost and, because other firms did not operate similar equipment, the cost often had to be absorbed by the company if it was to remain competitive. It was encouraged, however, that further waste treatment was likely to be made compulsory for all firms and this would raise the total cash spent on pollution control significantly. Management felt that there are significant economies of scale in pollution control in this industry and that stricter pollution laws could lead to mergers and possible closures.

Case Study 4

This establishment employs between 50 and 200 people and is owned by a large international company; no manufacturing processes are carried out at the plant but petrol and diesel engines are run for prolonged testing. The plant is therefore responsible for exhaust gases from both types of engine emitted from the equivalent of about 10 small trucks running 24 hours per day, 7 days a week. The plant is situated in open country and there is no pollution problem from this source; only one specious complaint had been raised by the neighbours about these exhaust gases. Large volumes of water are used for cooling the engines and this has a nitrite corrosion inhibitor added; every 24 hours a proportion of this water is released into the sewerage but the local authority made no charge for dealing with this effluent.

The plant obviously carries a stock of diesel and petrol and there is a danger of spillage of these fuels when tankers are unloading or during normal handling operations. The fuel stocks are enclosed in a protective bunker and to protect against accidental spillage the storm-water drains feed into a small lagoon where any oil can be quickly removed before it reaches the nearby stream. Every day the lagoon is checked for significant amounts of spillage but none has yet been reported and the firm had never been prosecuted for polluting the stream.

The firm used the local incinerator plant for disposal of waste paper which contains confidential matter and had not been able to find anyone willing to repulp it at a convenient distance. The firm felt that either the local authority could provide a "secure" paper collection service or small low-cost paper shredders could be developed.

The firm's most serious problem was that of noise. Diesel and petrol engines typically make a lot of noise and these are no exception. This presented a problem both within the factory and outside in terms of relations with the neighbours. Both aspects of this problem had been tackled vigorously; about seven years previously the management embarked on a programme to educate the workforce to the dangers of noise. Lectures from a consultant audiologist were arranged for those members of staff at risk, ear muffs were provided for all and poster campaigns devised to ensure that they were worn all the time. At first there had been little enthusiasm from the workers; they objected to the gibing comments of their fellows but eventually the "acceptance barrier" was passed. At the time of the visit most, if not all, the workforce at risk wore or carried their ear muffs constantly. The firm arranges six-monthly hearing tests for all staff that wish them and it was stated that two employees were showing signs of deafness. The firm had obviously done much to bring the dangers of noise to its employees' notice and to offer protection. The firm had not introduced any sanctions for workers who did not take advantage of the protection but was worried that, since legislation permits employees to seek damages for deafness, it may prove that employers are responsible even if the available protection is not used. Nevertheless, noise does not seem to be an important factor in determining labour turnover; some new recruits did leave very quickly but the rest stayed a long time.

Naturally, one aspect of the noise problem was in the sound insulation for the engines and for the buildings. Generally insulation of the engines was not possible for technical reasons but the exhausts are silenced. Sound absorption materials were provided inside the buildings and mufflers provided on air intakes and exhaust ventilators. Further, where possible, remote controls for the engines were being set up in a soundproof corridor but it would still be necessary for maintenance and assembly work to be carried out in very noisy conditions. Double glazing had been installed in all windows, double doors provided where necessary and, in the new buildings, the windows had been specially designed to minimise sound transmission. These and other measures helped to minimise the noise problem to neighbours and, in fact, the company had received about one complaint per year. The company claimed that it informed neighbours well in advance of work which could inconvenience and always followed up complaints quickly.

The plant more than satisfied the existing legal requirements but the company considered that existing legislation does not adequately protect people who live in rural areas. The respondent concluded : "if they can hear us at all, it is too loud".

Case Study 5

This factory employs nearly 400 people: another factory owned by the same firm a few miles away employs approximately 600 more. This company therefore was one of the largest we interviewed although it fell within our definition of "medium sized". The description of the plant and premises, however, relates to the foundry employing nearly 400 and making components for domestic central heating equipment.

The foundry operated three cupola furnaces, melting between 300 and 400 tons of iron per week; the operation of these furnaces led to considerable air pollution around the factory and complaints had been received from local residents who took the issue up with the local authority and their local MP. The firm had complied with existing local legislation regarding the operation of these furnaces and little could therefore be done; nevertheless the existing furnaces were in the throes of being replaced by electric ones. We were told the decision to replace the furnaces was solely an "economic" one but it did have the added advantage that the pollutants emitted by the new furnaces would be much reduced. In addition to the exhaust gases from the foundry processes, the firm also faced problems from the emission of sand and iron particles from the casting and fettling shops; these were removed, as far as possible, by extractors. The firm had also started the use of resin-bonded sand for moulds and this can give rise to obnoxious, though non-toxic, fumes. If this were to be continued on a larger scale, the firm envisaged installing plant to burn off these fumes before reaching the atmosphere.

Within the factory, face masks and dust extractors were used extensively; indeed there is a legal requirement, enforced by the Factory Inspector, for such equipment in certain parts of the factory. The running costs of this equipment were of the order of £1000 per annum and filtration plant for the new parts of the factory accounted for approximately one sixth of the total capital costs. The respondent suggested that employees were becoming more and more conscious of the problems of the environment in the factory and that the face masks provided were being used more and more; however, there had been no direct intervention from the trades unions at the plant.

By the nature of its production processes, the firm had no liquid effluent problems. About 20 million gallons of water were extracted from the river every year for cooling purposes and about 80% of it returns to the river - about 5 degrees warmer. The degreasing plant passed slighly alkaline effluent into the foul water sewer but the volumes were very small and the firm did not mention any effluent charge in connection with this.

Solid waste disposal constituted a major problem for the firm. Scrap metal waste was sold to merchants and presumably recycled, but by far the biggest problem was the disposal of about 300 tons of sand and slag from the furnaces every week. This was tipped on nearby sites by the firm and, in fact, there had been some difficulty over the disposal of the newer resin-bonded sands in the light of the 1972 Deposit of Poisonous Wastes Act; after local authority objections, the company had to change its tipping place. The company also tipped waste packaging material but the local authority was used to remove waste food from the canteen. The respondent saw this as the limit to which authorities should have to go since, although he would have liked them to do more, he recognised that the waste from this factory would, in fact, have swamped the existing local authority services. Solid waste disposal in terms of lorries, drivers and tipping charges cost the firm approximately £10,000 per year.

The firm also was faced by considerable noise problems within and outside the factory. The problems inside the factory were largely a result of the nature of the processes being carried on and little could be done about them; the company maintained a supply of ear muffs for those employees requesting them, although there was no compulsion. Problems outside the factory were a result of the operation of a second shift and ensuing complaints from local residents; the firm, through its trade association, had undertaken a study of the noise levels around the factory and certain items of equipment had been found to be responsible. Restrictions and specifications were being laid down relating to the new plant and the respondent estimated that silencers for the dust extractors added about 15 percent to their cost; the cost of ear defenders was no more than £1000 per annum.

Overall, despite the seriousness of its problems, the company spent less than one percent of its production costs on pollution control; nevertheless, the respondent was prepared to see heavier penalties provided "they are done fairly and squarely" - the respondent perceived that the Factory Inspector did not apply his standards consistently to all firms.

Case Study 6

This factory employs about 100 people and is owned by a large organisation in the dairy trade. Air pollution was not a large problem but liquid and solid wastes were, and clearly much thought had been put into these aspects at the design stage of the plant which was located on a new trading estate.

The liquid effluent problem was probably the most significant; firstly there was a problem of volume. Liquid effluent could reach as much as 270,000 gallons per day or nearly 25 percent of the sewerage plant throughput; the BOD content was approximately 730 ppm and pH between 6 and 8. The effluent consisted of dilute detergents and sterilising fluids but the source of most difficulty is the denatured protein and fats which are, in fact, biodegradable. When the factory was being established, the local authority did not have a sewerage system large enough to cope with this effluent and the firm was faced with the decision whether to build its own treatment plant on site. This would have had few additional advantages and would probably have incurred higher running costs than the current payments to the local authority; further, there was the likelihood that the scheme would have been vetoed by the local authority on the grounds that the offensive smell given off by such a treatment plant was undesirable in that part of the town. In the end, the firm collaborated with other (new) firms in the area and made a sizeable capital contribution to the cost of the necessary new treatment plant operated by the local authority. As a result of a design fault in the factory, the firm had been prosecuted for pollution of storm water drains but this design fault was quickly rectified, though there had been some loss of fish life in nearby rivers.

The solid waste disposal side was not so acute but had caused difficulties in the past. Trade waste was sold to farmers for feedstuffs but, apart from this, there was the usual packaging waste and food from the canteen. At first the firm had used the local authority service for this but their experience was so bad that they eventually resorted to a private contractor. However, negotiations were proceeding with this firm, in collaboration with others in the area, to try to persuade the local authority to institute some form of skip system of refuse disposal. These discussions had become bogged down because of the reorganisation of local councils. The respondent thought that this might be a profitable area of activity for local authorities, judging by the profits of the private contractors.

There were noise problems in certain parts of the factory and all operatives had been issued with ear protectors; these were hardly ever worn, however, and the respondent thought that personnel had not fully realised the long term consequences of not wearing them. Nevertheless, steps were being taken to attenuate the problem, soundproof booths were being installed and all new plant was being ordered to comply with specified noise limits.

Case Study 7

This small sized firm of 200 employees is part of the Vesty food group and has been engaged in felmongery for over 200 years, and much of the equipment has been installed for decades. Its chief product is sausage casings, whilst other byproducts include natural gut and sheepskins. The firm was a member of two industry associations, namely the UK Felmongers Association and the Association of Natural Sausage Skin Manufacturers. The main pollution problem that had plagued the firm for years was the problem of odours. This used to be bad but had been reduced to minimal levels. The reputation of the firm as a producer of bad smells still lingered in the vicinity.

The firm was not located in a smokeless zone but used oil fired heating and electrical power for efficiency reasons. Consequently small amounts of carbon monoxide and sulphur oxides were produced. Within the factory face masks were used to prevent inhalation of lime dust which was used for some cleaning purposes. Thus the real problem not covered by the Clean Air Acts the firm faced was the supposed obnoxious smell from the gut cleaning process. Before the war the gut used to be left hanging up for weeks in order that the insides could ferment away; this stank and the company tended to be blamed for any obnoxious smell in the town. Now, however, the gut cleaning process has been speeded up and consequently the smell problem has been greatly diminished. The firm's reputation had lived on, giving rise to continued public opposition and, as a result, one of the firm's offshoots in another town closed and moved to a new site in Doncaster; the smell continued though, since the old firm had been situated next to the local sewage works.

Liquid waste had always been a problem for the firm since a substantial amount of biodegradable animal wastes are washed out in the gut cleaning process. Sodium sulphide and lime are also waste products of the wool cleaning process. The Alkali Inspectorate made regular visits but most of the waste material was just washed down the drain to the main sewer. The firm paid £1360 per annum as a flat rate charge for using these facilities and an additional variable charge of about £1200 per year (dependent on the amount of effluent discharged). Its own monitoring and filtering system cost a further £3500 per annum in capital and variable costs. Some wastes are insoluble and these were screened off before reaching the main sewer and were taken away by a private contractor for dumping at sea. Twenty years ago all liquid effluent was discharged directly into the river but the pollution problem from this became excessive and the firm was forced to utilise the public sewerage and therefore substantially higher costs.

Solid wastes were a minor problem, but it is interesting to note that some animal wastes (skin, odd pieces of wool) which at one stage were thrown away were now sold and a profit of £2000 per annum made; this compared with the previous cost of £2000 to throw the waste away. Use was made of as much waste as possible; even the lime dust was used to fill in land on the factory site. Without the enterprise of this firm in disposing of its waste, the local authority would have been stretched to cope with many extra demands on its tipping space.

In total, pollution control accounted for between 5 to 10 percent of total production costs and it was likely to remain at this level for some time in the future.

Case Study 8

Up to six months before our visit the firm had two establishments, one on an industrial estate and the other near a large centre of urban housing. The firm dyes and bleaches lace and some nylon products. The latter establishment employing fifty men and having a turnover of £250,000 per annum had recently closed down and, according to the Managing Director, this was directly as a result of local resident pressure concerning fumes emitted from the chimneys. The problem in this case was fumes (mainly phenol) from the stentering process. The firm spent £8000 on boosting the chimneys in the plant that shut down but apparently this did not stop residents' protests and the firm took the drastic step of closing down.

The factory on the trading estate was relatively new but still had substantial pollution problems. The firm used £4000 worth of gas per month to heat the dyebaths and cleaning baths. The main fume problem, however, came from the dyeing and bleaching processes and was again concentrated around the stenter ovens. The firm had one 70 ft. high metal chimney which cost £5000. The respondent claimed that this was insufficient and that instead four of these chimneys were needed to make the production processes more efficient and not for the removal of an air pollutant which could be a public nuisance. (The alteration to chimneys was to let the gases escape through the factory roof.) We were told that the local authority claimed that three more chimneys would be an eyesore and that the firm must dispose of its fumes in another way. The Factory Inspectorate, on the other hand, had said that more chimneys were needed because the solution of letting the fumes leak from the stenter ovens within the factory and then out through the roof was detrimental to the health of the employees. To try to abate pressure from both sides, the firm installed £8000 worth of extraction equipment which aided the flow of fumes through the roof but was not dependent on a chimney, but this did not fully satisfy the Factory Inspectorate. Additionally, fume suppressants had been added to the bleach at a cost of £400 per year but with the effect of reducing the efficiency of the bleaching process. The alternative to all this was the installation of three machines, each costing £20,000, to actually destroy the stenter fumes but the firm felt that in the economic environment then pertaining this was more than the company could stand.

Like all dyeing companies, the firm has a liquid effluent problem. It had a throughput of water of $\frac{1}{2}$ million gallons per week and this cost some £1500 per year. Its effluent was not discharged direct into the sewers but rather went into a large drain in order that the acids and alkalis could neutralise each other. Even after the installation of this equipment (which cost £15,000) the firm still had to pay £2000 per annum in effluent charges based on the BOD, suspended solids and volumetric assessments. Biodegradable dyes were used in 95% of all cases.

The firm produced little solid waste, and what it did have it either sold or got a private contractor to remove at a cost of £500. The local authority had never offered a waste disposal service.

The factory had substantial noise problems from winches and live steam injection. To eliminate this the firm had embarked on a policy of machine enclosure which would cost £4000 and a slowing down of motors, reducing efficiency slightly. Generally the firm felt that since it budgeted for pollution control measures yearly at £10,000 (which represented 2% of turnover and 20% of profits) then there should be some form of government assistance as they were the ones imposing the controls.

Case Study 9

This textile firm of 900 employees is situated in the north of the region. The firm has been established since the late 18th century and employs mostly local female labour. The factory is two miles from the nearest urban area and is situated by a small lake and river which used to form the basis for its power supply, but into which it now discharges effluent and from which it takes water for all uses except drinking. The factory is not connected to the main sewers, had little or no contact with the local authority and all effluent processing and dumping occurred within the factory grounds (120 acres).

Electricity and coal formed the main source of heating and power. The dyehouse process used 110 tons of coal per week; before the smokeless zone came into force the firm used a lot more coal and polluted the atmosphere to a much wider extent. However, when the smokeless zone was introduced the firm planned its combustion processes in liaison with the local health inspector and this had resulted in an agreed level of air pollution. (Since the factory was so far from human habitation the problem had never been really acute.)

Within the confines of the factory the major problem had been cotton dust. This was not particularly bad at the time of the visit but poor grade cotton could cause problems. First discovered in 1920, it is now apparent that cotton dust causes a respiratory disease, byssonossis, which may take up to 40 years to reveal itself. In this factory only 20 people who operated the cotton opening and cording were at risk but face masks and dust extractors were used, especially when cleaning machinery. Unfortunately the workers did not always make adequate use of the facilities and were compensated by means of muck money for any such work. Such problems do not exist with man-made fibres, and in relation to the Lancashire cotton mills the problem was relatively minor. The Factory Inspectorate, management and trades unions had been cooperating since the 1920s to reduce the risks from cotton dust.

Liquid waste resulted largely from the dyeing process, most dyes are biodegradable over a long enough period of time. The firm had worked in close liaison with the Trent River Authority setting up a chemical treatment and filtration plant to remove the toxic substances. The suspended solids are precipitated off and the remaining liquid discharged after treatment into the adjoining river. Recently £16,000 had been spent on a new filtration plant, the running costs of this and the pre-existing facilities were expected to be about £5000 per year. The new plant was installed in order to cut down the risk of accidental pollution and speed up the effluent treatment process, in particular by removing the element of human error since it had an automatic maintenance system.

There were two main types of solid waste; firstly the precipitated sludge from the dyehouse ($1\frac{1}{2}$ tons of this are produced per day); and secondly fibre waste from the production processes. The latter was wholly recycled in the factory unless the quality was too low when it was sold to other firms. The sludge had been dumped on the factory premises and the firm thought that in the long run the dumping of this waste could present a serious problem; it is not poisonous but is unsightly. Dumping costs £2000 per annum but the cost would be much higher if it had to be tipped elsewhere.

The noise in the factory had been cut to an absolute minimum with the use of fibreglass cladding and oil baths, before then the bad noise had a recognisable effect on the workers' efficiency. At the time of our visit only one part of the factory had a noise level in excess of 90 dBA (the proposed limit set by the government) and steps were being taken to reduce this. (The costs of changeover were not relevant since primarily the changes were undertaken to promote increased machinery efficiency.)

Overall the firm spent less than 5 percent of total production costs on pollution control. Its parent company had a definite corporate policy on environmental problems and this meant much of the improvement inside and out of the factory was self-generated. It was the opinion of the management that the authorities concerned were too slack and did not have high enough standards on the question of pollution control. The respondent felt that many other local firms in the area were polluting the district and not subject to sufficiently stringent controls.

Case Study 10

This firm of dyers and finishers was concerned about the highly alkaline nature of its trade effluent resulting from the mercerising processes undertaken. In fact the problem had at one stage been sufficiently acute for the local authority to consider seriously closing the plant. Eventually the firm managed to devise and develop plant of an entirely novel form in order to neutralise the alkaline wastes and, at the same time, one additional but non-critical pollution problem was overcome.

The effluent from the mercerising process is sprayed from the top of a tower through which flue gases from the water boilers are passed. This self-buffering system therefore results both in safer trade effluent and cleaner air around the factory since the alkaline effluent neutralises the acidic flue gases. The capital cost of this equipment was £7500 in 1967 and the current operating costs were of the order of several hundred pounds per annum (this was mainly for an electric fan to pull exhaust fumes through the tower). The performance of this equipment had been so successful that the firm was marketing it, in addition to its dyeing and finishing activities.

Case Study 11

This firm has 1000 employees and is by far the largest establishment in our sample, although it is NOT part of any integrated grouping. The firm is a genuine independent of that rare breed which is large enough to compete in terms of modern equipment with the subsidiaries of firms like Courtaulds and Carrington Viyella.

The dyeing operations are strictly on a commission basis for the finishing of all types of cotton and synthetic knitted materials. The firm is a major water user and effluent discharger with a total throughput of 4½ million gallons per week, most of which is abstracted from the nearby river, and then passed through a water softener. The firm carries out partial pretreatment of its effluent but lets the sewage works cope with the bulk of treatment. Indeed, it was the proximity of the sewage works and river which was a major locational factor for the firm. Clearly then the major feature of this firm is its very large throughput of effluent, which as of yet it does not consider worthwhile treating completely itself. The firm already softens its water input with a plant purchased six years ago for £150,000. The total cost of water is 10p per gallon which is relatively cheap and makes it difficult for the firm to consider changing location and reduces any stimulus to use water-saving techniques in dyeing and finishing. A variety of dyeing techniques are used, some water-intensive like the winch machines (with a 25 : 1 liquor ratio) and some less so like jet dyeing (with 5 : 1 liquor ratio), and solvent dyeing.

The liquid effluent from the dyehouse is treated in two ways prior to discharge. Firstly, from the jet dyeing machines, the company uses heat exchangers to reduce the temperature which would be in excess of consent conditions if discharged. This heat exchanger in fact pays for itself (although the firm has an excess of hot water) but is an essential item of equipment since the company would have to reduce the temperature of its effluent from the jet machines in some other way if it had no heat exchangers. It is, however, at something of an advantage being a large unit since it has enough cold effluent from rinsers on other machines that are simultaneously available to cool the effluent from the jet machines if necessary. The firm has a small lime treatment plant but nothing for equalisation of effluents on any scale, since they have a sufficiently heterogeneous flow of effluent to equalise itself on the way to the sewer. This clearly is a benefit only the very large firms can derive. They have also installed a stockdale filter to reduce suspended solids to below 400 ppm. This cost £5000 and has only a small recurring maintenance bill. The effluent is then discharged to the sewer and, with the new Severn Trent water charges, this will cost £50,000 per annum. The firm is also operating under a mass discharge constraint and has to reduce its volume of effluent discharge by about 12 percent in the next 12 months.

Thus whilst the firm has much up-to-date equipment, it has only recently begun to seriously investigate water conservation. Continuous processes are a way other large firms can partly solve this problem but this is because they are normally part of an integrated group and treat their own material with some homogeneity of quality. The firm has approached six main suppliers of end-of-line equipment which would recycle water and reduce total water throughput but has found that the technology of water re-use is still not good enough to recycle water in a dyeworks. They have thus turned to process changes. The immediate alternative was to use fewer baths and turn the taps down; this can save 10-20 percent of water according to a recent industry association survey. Scour/dye chemicals have also been tried and do work but are slightly more expensive. Solvent dyeing has been considered but is not possible on any large scale because of the poor range of dye colours available. Thus the company still uses substantial quantities of water and will require some major innovations if this is to be altered.

The firm also have some slight problems with stenter fumes which killed off some local vegetations. However, a zero cost change of dye carrier largely solved this problem.

Internal health and safety requires little expenditure, apart from the mandatory dust extractors, face masks and ear muffs in the dye store and the softening plant.

Generally this firm should be able to cope technically and economically with most pollution problems; its sole disadvantage to other large firms in integrated groups is that it has not quite the same technical resources and cannot use the continuous processes they can use. Even so, the problem of water use and effluent discharge has, combined with the revised charges, become a substantial expenditure and this has forced the firm to rethink on reducing liquor : material ratios.

Case Study 12

This company has a total of 1000 employees, of which only 200 work in the dyehouse. The company is a subsidiary of a large vertically integrated group and, because of this, over two-thirds of its work is of a continuing nature from the knitting operation (mainly synthetics). The respondent in this firm continually stressed the advantages as he saw them of being a part of a larger group. In particular he saw the use of the holding company's finance for research into water use and heat conservation as very important. Their only problem in the path towards more efficient dyeing and reduced pollution control costs was the cramped position of the dyehouse which was not very suitable for major movements of process equipment. However, this particular company had tried out many different dyeing techniques. Its weekly effluent discharge at present was 1.5 million gallons, for which their new charges would be slightly in excess of £2500 per annum. For the size of firm and throughput of water this figure is very small and it reflects the considerable advance that the integrated firm with substantial parent company finance has been able to make. Nevertheless, the firm's sister company pays no effluent charges at all and this clearly demonstrates how charges can vary substantially within and between regions.

Extensive heat recovery systems have been installed for recycling of heat and abatement of warm effluent. The firm, however, does not use any equalisation process for the effluent other than the mixing that goes on in the sewer; this is satisfactory because of the large and heterogeneous volumes of effluents that are being discharged all the time. The most important reductions in liquid effluents have occurred because the company has introduced continuous rather than batch production (something only an integrated firm can do). At its simplest, this means that production scheduling can minimise water and chemical input because of the known parameters of the process and economies in water use through the existence of linked processes. This was particularly so for the continuous bleaching of whites.

The company was also making greater utilisation of fast processes such as jet dyeing, which had a much lower liquor ratio but in fact were installed because of speed of operation, and particular suitability to polyester rather than for effluent consideration. They were also using a solvent dyeing machine but the capital costs of this were very high, necessitating excessively expensive dyestuffs. There also appeared to be a health and safety risk involved because the dyeing was in perchlorethylene which reputedly had possible carcinogenic effects. Thus in this case there was a possible trade-off between high-volume low-danger effluent for another low volume but potentially much more dangerous one.

The company had few air pollution problems with the exception of a case about a year previously when local residents complained at the smells coming from the stenter when sulphurous dyes had been used. The company changed to vat dyes and the complaints subsided, although the cost in dyes to the firm increased four-fold.

Generally, though, the firm was free of pollution problems and was very much an innovator in the industry. Both of these facts seem dependent on the firm's position within a large textile group. The only really severe costs that could impinge on the firm's operations were if pollution control demanded any more space; if this was so then the firm would be facing very serious problems.

Case Study 13

This company is an established firm in the region and is now a member of a large national integrated group. It employs approximately 250 people in the dyehouse and works mainly on a non-commission basis, although with the recession firms had begun competing in the commission dyeing area as well. The firm concentrates on dyeing synthetics and it is planned for it to become one of the main centres of the group's operations. The water usage is in the range of 1.5 million gallons per week and its charges amount to £18,000 per year for effluent discharged untreated direct to a sewer for which the company made an initial contribution of £10,000. The firm is relatively unconcerned over the effluent charges, although admits that the increases will be very difficult to pass on to its customers because of the depressed state of the industry (although because they were operating at the high quality end of the market, there seemed considerable margins already), and the degree of price competitiveness that the East Midlands section of the textile industry in particular had encountered. The firm used some low liquor ratio techniques, similar beam and high temperature dyeing but commented that the quality and shade of dye was the major factor affecting quality of effluent. In particular light shades needed fewer rinses and with beam dyeing the acid dyes became quickly exhausted; however, for dark shades a lot of rinsing was needed as this added considerably to the effluent load. So far the firm had gone no further than investigating new dyeing techniques (pressure beam machines) with lower liquor ratios and, even in these cases, the objective was to stop the use of carriers that were necessary for dyeing synthetics on traditional machines. They also felt considerable effluent savings were made because 95 percent of their business was continuous dyeing which was just more efficient in the use of all factors of production.

Case Study 14

This firm was the smallest in our sample of dyers and finishers, employing 15 workers, and it dyes pantihose. The problem of water charges and effluent costs seemed important to our respondent who said that he felt "... the river authority was trying to put small firms out of business" although he went on that the pressures of running a small firm were such that maybe they would "do him a favour if they did put him out of business". The firm uses 17,000 gallons of water per week and, last year, paid effluent charges of £1000. The respondent thought this could be five times as much next year. The firm does not treat the effluent, has made no in-plant changes to the process and still uses some antiquated paddle machines. Whilst the firm felt the increased charges were important, they also felt that the large integrated groups were making life difficult for the small firm in any case. In order to reduce water and effluent charges an informal approach had been made to the firm to participate in a joint reclamation project; they had had to reject the idea, though, because of cash flow problems. Generally the firm accepted the need to control pollution but thought that the cost would have to be counted in the exit from the industry of many small independents such as themselves.

Case Study 15

This company employs 100 workers and has spent a good deal of cash on new investment in the last five years. It has bought seven jet dyeing machines (costing £20,000 each) mainly for the dyeing of knitted fabrics. Its effluent costs are £2500 per year and generally the firm seemed totally unconcerned about effluent and water usage, which is probably a direct result of the firm having up-to-date low liquor ratio dye machines.

Instead, the firm was presented with a fume problem from its stenters. The firm had bought for £10,000 a catalytic after-burner to put on its stenters after the local authority had advised it on this course of action because of complaints from local residents. Unfortunately the type of material causing the problem in the stenters was fluffy and constantly set the after-burner on fire. The firm had thus had to stop using this and instead was now considering putting an £8000 steel chimney over the stenter for high dispersal. Until it does this, they try only to use the stenter when the wind is in the right direction. Substitution of dyes and chemicals had been tried but none had worked. The firm felt it was in a hole even though it had done all the local authority had asked. This was clearly a case where available technology had not caught up with the specific nature of the pollution problem.

Case Study 16

This foundry has a labour force of 250 and is part of a large horizontally integrated engineering group. The foundry produces parts for the marine and automotive industry on a production basis. The capacity of the plant is substantial with three 10-ton arc furnaces and one 6-ton coreless induction furnace. The firm is not located in a smoke control zone. The pollution generated from these furnaces is a lot less than that from a coke-fired cupola but there are still substantial fume problems inside and outside of the factory. The electric arc furnace comes under the auspices of the Alkali Inspector.

In order to ameliorate the pollution to the external environment the firm has installed a wet gas scrubber which cost £35,000 in 1973 and, when working, achieves a high degree of efficiency. However, it is subject to frequent breakdowns and this has caused considerable production problems at the plant, although the plant has never actually closed because of its uncleaned emissions. The cleaned gas is emitted from an 80ft. stack but the residues from cleaning result in a highly viscous waste amounting to 5000 gallons per week, most of which is discharged down the sewers after filtration, the remaining sludge being taken away by a contractor at a cost of £3200 per year.

There are also treatment fumes from the desulphurisation process and these cause especially severe problems within the workplace, as well as the usual dusts from grinding, fettling and shotblasting. A sophisticated dust extraction system is used in conjunction with bag filters which prevent the ventilation air either causing problems inside or outside the factory. The total cost of this system is estimated at £200,000.

Solid wastes are only a minor problem since no sand moulding is used; however there is approximately 20 tons of refractory lining waste per day but no cost data is available for this since its removal is by part of a contract for the whole group.

There are severe noise problems in the factory but so far only traditional means of protection have been used, and there are no noise problems to the external environment. Because of the internal fume and noise problems, labour turnover is high even though the firm pays wages 20% higher than the national average earnings.

This firm uses melting techniques normally associated with reduced pollution loads but has found that considerable expenditures were still necessary and large problems still existed in the internal environment.

Case Study 17

This firm has a labour force of 100 and makes a large variety of castings. It is an independent jobbing founder, situated in the north of the region and comes under the auspices of a local authority who, according to our research, have considerable expertise, resources and motivation in pollution prevention.

Two 5-ton coke-fired cold blast cupolas are used and a wet arrester is fitted to prevent dust and fume reaching the atmosphere. The firm is not in a smoke control zone. As well as wet arrestment there are also considerable internal controls to collect fettling fume and dust via extractor fans and bag filters, although airborne emissions from the moulding shop are extracted direct to atmosphere.

The use of a wet gas scrubber was made after several meetings with the local authority who had had several complaints from residents. The liquid waste from the wet arrester is discharged direct to the sewer at minimal cost. The company has specialised within the founding industry and wants to expand its operations, although it estimates that £100,000 will have to be spent on new wet gas cleaning plant.

Substantial solid wastes result from the process as 100 tons of sand are used per day; at present none of this is recycled but since it now costs the firm £20,000 per year to dump it a sand reclamation plant is now being considered.

Noise is also a problem but only minor measures have so far been taken. More will soon have to be done as the company is now encountering complaints from local residents.

We found this firm extremely knowledgeable about legislation and techniques of control. They were also profitable by jobbing foundry standards yet could still only make a 6% rate of return on capital. The firm's total pollution control costs were about 10% of value added and it was envisaged that these would rise still further in the future. Management considered that many of their competitors had found the burden of pollution control too much on top of their declining profit margins and had left the industry.

Case Study 18

This firm is an independent jobbing founder situated near an urban area. It employs 48 workers and has two old-fashioned cupolas. One melts 3 tons and the other 4 tons per hour. There is no arrestment device on the former and only an inefficient dry arrester on the latter.

In view of the new legislation, the firm had decided to cut its capacity and scrap the smaller cupola when forced to and also estimated that £10,000 would be needed to bring the other cupola to the required standard. At present this was beyond the firm's resources but it hoped to be able to borrow the money. One problem that the firm considered important was that, as with its in-factory extraction equipment, there were substantial economies of scale to be generated in pollution control and that small firms frequently had to buy equipment which was far too large for their uses. In this particular case it was claimed that output could be doubled with no extra pollution control cost.

Solid waste in the form of sand is dumped at a cost of £2000 per annum; this is causing problems at the moment because the local authority claims it is interfering with drainage in the area. So far no other alternative sites have been suggested.

The firm operates in a market segment which is especially prone to price cutting and feels that all costs of pollution control would have to be absorbed.

Case Study 19

This firm is situated in a large urban area in the north of the region. It employs 150 workers and makes high grade castings and some ductile products. An important feature of this case was that in 1971 it changed from cold blast cupolas to electric induction furnaces, one with a 2-ton capacity and the other 5-ton. These were installed because (a) the firm could control the quality and the emissions more easily, (b) it could more easily produce ductile iron, (c) it was situated near to houses and recognised the potential problems that would be faced if it continued to use coke-fired furnaces, and (d) it had investigated the cost of a wet scrubber for the old plant but found it expensive and not too easily compatible with the plant. Thus emission control was only one reason for the switchover. Most of these expectations proved to be right. The external pollution problem was solved and the firm is now able to make better quality castings where demand elasticity is less. However it does now have a fume problem inside the factory and already £30,000 has been spent on dust extraction equipment.

The firm has no liquid effluent problem but has solid wastes in the form of used sand; the cost of this tipping had risen steadily and was now £15,000 per annum. However this was a minor problem compared to the one of actually finding new tipping space as the local authority had now banned the waste from their tips. These pressures had led the firm to buy a sand reclamation plant which, like many pollution control and recycling devices, has been bugged by technical problems.

Noise was a problem but this was a traditional foundry area and the labour force had a reasonably high endurance threshold.

At present pollution control was less than 5% of value added and in-plant changes in the form of new furnaces had proved to have commercial benefits outweighing their extra cost. However, the firm thought controls would become stiffer, especially on the fume from the electric furnaces which had at one time been considered harmless but were now subject to additional scrutiny. Even so, the firm had specialised so much that they felt they could pass on these costs.

Case Study 20

This is a medium sized establishment employing 100 men and is part of a larger engineering group. The respondent considered that only because of cross-subsidisation by the group to its foundry operations had this particular firm survived. There was one 2-ton cupola and two 5-ton cupolas. These were all coke-fired, cold blast and were all fitted with wet arresters at a cost of £20,000 three years ago. The firm was in a smokeless zone and there had been considerable local pressure to abate air pollution. It was felt that as founding was not traditional in this area the firm had come in for a lot more pressure than its competitors in other areas. The council was still not satisfied with the measures taken and the firm was about to purchase a supplementary firing system at a cost of £20,000 which will reduce fuel inputs as well as pollution. It had been pressure from the local authority that had forced this measure but, following tests in conjunction with BCIRA, it felt that it could amortise itself in three years by fuel saving and was in fact quite happy that it had been forced into this position.

As with all foundries, there were health and safety problems but in this case these had only necessitated at £15,000 expenditure.

Similarly liquid waste costs from the gas scrubber were negligible. The firm used green sand for moulding which, whilst more expensive, had the added benefit of being easily cleaned without sophisticated equipment and thus the firm only incurred £6000 in dumping costs for solid waste.

Noise was not an external problem but caused considerable trouble within the factory. So far not a great deal had been spent but there had been considerable costs through labour disputes and stoppages concerning internal working conditions. This latter point is particularly interesting because, in the more traditional foundry areas, internal conditions proved to be much less of a problem because of apparently different sub-culture thresholds.

The firm felt that environmental control was now a major problem and it was considering replacing one of its cupolas with an electric induction furnace which it envisaged would present fewer pollution problems because of the greater ease of control and would also produce a better quality product. Probably its greatest complaint was the unequal application of controls in this industry.

Case Study 21

This firm is a large foundry employing 700 men. It produces high quality grey iron castings and is part of a large automotive group. It has substantial melting capacity in the form of two 6-ton and two 10-ton cold blast cupolas. Unlike many foundries, these work continuously five days a week and 8 hours a day. Additionally there is an electric induction furnace with a 2-ton melt rate. The potential pollution problem is thus large. Black smoke, dust and metallurgical fume (and oily smoke when using the induction furnace) are the main airborne emissions.

However, the only pollution controls are a series of extractor fans to disperse these pollutants from the chimney without any form of cleaning. The firm clearly caused a lot of pollution but apparently no-one seemed to worry because its geographical location meant that only a few households were affected. Its relations with the local authority were described as "excellent". Even so, the firm felt it necessary to control itself and was now considering a revolutionary design of hot blast cupola which would greatly reduce emissions although the initial motivating force behind this was for quality control and product improvement.

Within the factory conditions were excellent and management estimated that its internal control equipment cost £750,000. These high class working conditions were also matched with above average labour force earnings and we would conclude on this evidence and that of our other interviews that there is no trade-off between earnings and conditions of work inter firm. These factors appeared to be complimentary rather than substitutable benefit factors.

Small quantities of liquid effluent were produced and these were discharged direct to the sewers at minimal cost. As with all foundries, there were large amounts of solid waste and in this case it was dumped when too contaminated. Approximately 80 tons of sand are dumped each day on local authority tips free of charge, the company paying only £10,000 per year in transport costs. Noise was a problem but was solved adequately with small investments in ear muffs and soundproof booths.

Generally the firm had substantial technical and financial resources and yet had made only minor moves on the pollution front. Internally it had invested very substantial amounts after pressure from trades unions but there appeared no similar guardian of the external environment.

Case Study 22

This firm has approximately 150 employees and produces specialist equipment for a consumer market in which it has a considerable brand following. The firm is reasonably profitable and is at present being taken over by a large national engineering group. It has three plants and considerable excess capacity at present. At the plant we visited there were three cupolas, one with a 5-ton/hour and the others with 3-ton/hour capacity.

Wet arresters were used to prevent emissions to the atmosphere but they were causing a corrosion problem and the liquid waste was considerably polluting the nearby stream although, as yet, there had not been any official complaints. Whilst the firm had considerable technical expertise, it was unable to estimate the gas cleaning efficiency of its equipment but considered that still too much fume was being emitted. It was thus considering some new plant which had a pollution control component of £60,000 and would at the same time shut down one of its other plants in order to get full use from this investment.

Internal fume conditions were dealt with in the normal way and this had so far cost £20,000. Solid wastes, too, proved minor as the firm had its own tip and had to pay only transport costs. Noise controls amounted to only £1000.

Thus the firm had encountered only minor pollution control costs but envisaged these would rise substantially in the future and, even in this specialised market, would result in lost jobs.

Chapter IX

INDUSTRY'S ATTITUDES AND INFORMATION ABOUT POLLUTION AND COSTS OF CONTROL

Whilst pollution and pollution control are objectively quantifiable, we felt it important to analyse the more subjective area of attitudes and stock of information that our respondents had towards environmental matters. Thus in the final section of our questionnaire we asked firms about a variety of topics concerning their attitudes to existing controls and standards, the perceived impact on total costs to their firm of pollution control expenditure and the possible need for public subsidies. The results we have are based on simple questions and we make no pretence to the scientific validity of our analysis of replies since many firms preferred not to answer questions in this latter section; however, we feel our results are worth documenting since some important inferences may be able to be made.

Replies from Firms

Data on pollution control costs have already been dealt with in appropriate sections of this report but in Section III of our questionnaire we asked about the perceived total cost of pollution control as a proportion of total production costs. Rather than a strict documentation, this question was designed to elucidate respondents' attitudes and information on costs. Towards the end of the research we realised that this may not be a particularly important measure of the impact of controls on industry and we enlarge on this point later in Chapter X.

Nevertheless, our analysis of the proportion of total production costs that pollution control accounted for is shown in Table 9.1. The foundries were clearly the hardest hit in terms of tangible and recognisable costs and in the random sample two of the three firms in category two were engaged in metal manufacture, yet according to our research they were also least affected by in-process changes which instead had been particularly prevalent in

TABLE 9.1 Perceived Pollution Control Costs as a Proportion of Total Production Costs

	Less than 5%	5-10%	10-20%	More than 20%	No Reply
Random sample	31	3	0	0	28
Founders	18	7	1	0	3
Dyers and finishers	25	1	0	0	6

textile dyeing and finishing. That the high costs seemed clearly related to airborne emissions is a possible indictment on our system of control which tends to be concerned with the immediately recognisable and easily perceived form of pollution, which may not be the most dangerous.

Table 9.2 records whether the firms felt the costs of pollution control would change significantly in the future. The foundries were clearly worried about future developments, probably rightly so since our evidence in previous chapters suggests that new pollution control is correlated with previous pollution control as well as the level of pollution, a fact

TABLE 9.2 Expectations of Changes in Pollution Control Costs

	Likely to Change	Unlikely to Change	No Reply
Random sample	12	23	27
Founders	23	5	2
Dyers and finishers	12	8	12

which we feel illustrates how information is central to the whole environmental issue. Rather surprisingly, the textile dyers and finishers did not unanimously think think that higher control costs were imminent, even though the river authorities were quite explicit about increasing effluent charges and local authorities were becoming more concerned about stenter fumes. It was noticeable that those firms in our random sample which considered costs likely to rise often did so because they were most concerned about noise control. We considered as important one firm's belief that in founding changes in process technology could naturally alleviate the pollution problems but at a high cost through a switch to electric induction furnaces from standard cupolas. We also received many emotional but unquantified comments about possible changes in costs arising through total commitment to control. Some firms were able to specify the extent to which their pollution cost commitment was expected to change; one firm was budgeting for increased capital expenditure of £2½ million over the next five years. This was an exception though, and of the thirteen firms which were able to estimate cost changes, nine were from our sample of founders which anticipated increases of the order of 100 percent or so. It was our impression that these conjectures about future changes could adversely affect investment plans for future productive plant and equipment.

We also asked the firms how they dealt with pollution control costs and, in particular, whether they absorb the costs, pass them on to consumers in the form of higher prices or reduce the size of their operations and cause loss of employment. There was much evidence of full-cost thinking as many of our respondents felt that all overheads would automatically have to be passed on as a matter of course. However, as Tables 9.3 and 9.4 show, there were substantial differences between the various groups of firms in their response to these questions. In particular, the founders felt that the costs would have to

TABLE 9.3 The Incidence of Pollution Control Costs

	Absorb	Pass on in Higher Price	No Reply
Random sample	21	6	35
Foundries	6	14	10
Dyers and finishers	16	7	9

be passed on in higher prices, but that in the long run this would inevitably lead to reduced employment opportunities in the industry. This long run view of the impact of control costs via higher prices to reduced activity is consistent with the lack of immediate foreign

TABLE 9.4 Anticipated Employment Effects of Pollution Control

	Reduced Employment	No Effect	No Reply
Random sample	4	21	37
Foundries	11	10	9
Dyers and finishers	11	13	8

substitutes for most foundered products and the general low profitability in the industry. This contrasts sharply with the replies of the textile dyers and finishers, most of whom felt that the costs would have to be absorbed and often they made explicit reference to the monopoly buying power of their chain store customers and the growth of cheap imported textiles as factors which prevented them passing on pollution costs in the form of higher prices.

One firm in textile dyeing and finishing commented that the industry should be able to absorb the costs by operating its processes more efficiently; whilst we had evidence that firms were increasingly concerned with this aspect of good housekeeping, this was the only instance where a firm volunteered it to be a result of pollution control. Some founders had already cut down their labour force and were able to point to other firms going out of business; we found this happening to a lesser extent in dyeing and finishing.

Most firms in textile dyeing and finishing and founding felt there was a strong case for government assistance with pollution control expenditures; indeed, one firm saw it as the only factor that would prevent large scale closures. Only a small number of firms in our random sample concurred with this view. Respondents were also asked for their personal rather than corporate view of the existing standards and controls of pollution and the replies are tabulated in Tables 9.5 and 9.6.

TABLE 9.5 Attitudes to Monitoring Pollution

	Too Slack	Adequate	Too Tight	No Reply
Random sample	22	15	0	25
Foundries	5	16	0	9
Dyers and finishers	8	18	1	5

Clearly those respondents in our two worst hit industries feel happier about monitoring by local authorities and central government than firms in general, but it is noticeable that only one firm feels that government monitoring is too strict. Table 9.6 confirms the personal view that pollution controls are far from being unwarranted and in general are adequate. Surprisingly most respondents in the random sample, and a substantial number in the other two samples, felt that present penalties were inadequate. Ninety-two percent of those in the random sample, 50 percent of those in foundries and 30 percent of those in textile dyeing and finishing answered that they were prepared to see heavier penalties introduced, although one respondent made the very reasonable point that firms need help and advice, not legal enforcement, whilst in any case others felt that few policing arrangements could achieve anywhere near full efficiency of detection.

TABLE 9.6 Respondents' Perception of Existing Standards

	Too Slack	Adequate	Too Tight	No Reply
Random sample	20	17	0	25
Foundries	0	19	2	9
Dyers and finishers	2	24	1	5

In general the often qualitative replies to these various attitudinal questions help support our general thesis that pollution control costs are heavy only in specific industries but that even in those industries firms frequently are not only uncertain about the extent and effect of future legislation but are also unaware of some of the existing hidden costs of pollution control. The state and nature of the controls are seen as fair and adequate and any grievances that are felt could be alleviated by short term loans and government technical assistance. It is clear that the interaction between policy makers and firms is not considered adequate and that greater liaison between the two would greatly facilitate the implementation of controls that are generally regarded as fair.

Finally we were able to make some subjective impressions about the attitudes and information our respondents both in the firms and local authorities had towards problems of the environment in general and their own operations in particular. In the firms there seemed a very genuine concern about environmental matters, but it was our impression that in many cases respondents were confused about various controls and, indeed, occasionally sought our advice as to the current state of play. To a certain extent we felt that where there was ignorance of pollution and pollution-related topics such as recycling it was because the problem until recently was relatively minor to many firms and they had never really thought about it because it impinged little on their activities. Many small firms cannot afford specialist personnel and to some of them even simple technical relationships which help define pollution could be confusing. On the other hand, there was a small minority of firms which had often been in close contact with local authorities over pollution which felt that pollution control was another aspect of the state's take-over of private enterprise; this was not always because of the physical controls a firm had to comply with but was often a function of the number of contacts made and forms to be filled in that took up the time of the small entrepreneur. In firms with under 50 employees the managing director might well perform many different functions in the firm and to him the time taken on environmental matters may constitute a severe nuisance.

Local Authorities

Our local and central authority respondents were mainly quite conscious of the threat to industry and the economic health of the country of too much interference on environmental matters. This seemed especially so when local pressure groups used their influence to affect the council's operations. For example, we found that those areas which contained coal mines had few smoke control zones. One of our respondents in a local authority commented that "... essentially pollution control is more concerned with the cooperation and education of industry than with legal action", but the same respondent continued that without a great deal more scientific expertise and resources, local authorities would be unable to cover the vast heterogeneous range of pollution problems encountered.

Our impression was that at the local level councils were highly susceptible to pressure exerted by local industry and were all too aware of the economic consequences of severe environmental enforcement. They were also acutely aware that, in many cases, tracing the source of pollution was a difficult and emotive problem. We reported in our random sample the case of the fellmonger who stopped operations in one town but whose departure failed to stop the odours previously attributed to it, which eventually turned out to arise from the local sewage works. Monitoring difficulties are thus prejudicial to effective control. One foundry in Leicester was frequently accused by local residents of polluting the area and it had to secure the services of the industry research association, the manufacturers of its dust arrestment equipment and the local authority before the problem was traced to another source. With rapid changes in process technology it seems likely that the problems of tracing the source of pollution and investigating the harm it causes will become more and more difficult.

The approach to pollution control seems to use prosecution only as the last recourse. In only four cases in the last twelve months did we find that there had been legal action and the use of abatement notices was also rare. Formal hearings and court action demand too large a proportion of the time of the scarce manpower that does exist.

Local authorities, in our view, tackled pollution in an essentially pragmatic way and this was partly through lack of resources and partly through their concern for the local economy. Since the water authorities are outside the scope of local politics, it will be interesting to see what stance they take and how they tackle any problems that arise.

Chapter X

CONCLUSIONS

As we have said in our introduction to this report, we believe this work to be the first systematic attempt to study the pollution control costs of British firms, especially those in the small and medium sized categories. We have surveyed the four main areas of pollution - those of the air, water, solids and noise; the only area we have omitted is that of visual pollution but since we largely believe this to be a problem for large scale industry we feel justified in its omission. Nevertheless, in the two cases where we did encounter this issue, one firm had devoted a great deal of its resources in combating the problem whilst in Case Study 8 we referred to a smaller firm that was unable to implement its chosen pollution control technology (high dispersal of stenter fumes) because local authority planners felt that a chimney would represent a blot on the landscape. Visual pollution control in the context of the small and medium sized firm is likely to be represented by these indirect costs which are almost impossible to estimate.

In air and water, industry contributes less than 50 percent of the volume of pollution and we find that in these areas the small and medium sized firms contribute a small part of the industrial pollution load with the exception, perhaps, of textile dyeing and finishing where the volume of effluent generated is high even in small firms but where its potential harmfulness is slight. There are, of course, localised pockets of high industrial pollution, typically where heavy industry tends to be concentrated. In the case of solid wastes, however, industry generates more pollution than households but the contribution of small and medium sized firms is relatively slight. We also believe that noise problems do not principally arise from industry but again the large firms have the greatest problems.

Defining the Pollution Problem

The functional relationship between the output of pollution and the harm done to the environment is not a simple one for many reasons. Firstly we do not always know the eventual long term impact of many pollutants. Secondly society's attitude to the perceived damage caused is itself variable. Typically we found that in traditional industrial areas pollution was an accepted fact of life with which the local population had apparently come to terms, and we have some rather sparse evidence that complaints to local authorities about industrial pollution are more likely to come from new residents than older ones. As we pointed out in our chapter on the economics of pollution, it is not always an absolutely clear cut case that industry should be controlled when residents have a choice of where they live and this is especially so when industry was located in an area first. However, as society's expectations increase and change, we might expect little sympathy with this view, although we believe it one that should not be forgotten. Generally, then, pollution outputs can be measured objectively but the subjective judgment of the community is just as important in assessing the extent to which pollution should be controlled. Inside the workplace this variability in individual sensitivity to internal pollutants was also evident, as we pointed out with particular reference to the foundry industry.

Even though we can measure the output of pollutants in objective terms, it does not mean that the harm caused to the environment is a linear function of the output of pollution as judged by the quantity of dust, grit or smell. In particular, we found many instances of small firms situated in either small urban or rural locations that emitted pollutants but caused little concern because the quantities involved were not substantial enough. At the other end of the spectrum, there were small localised pockets of industry where the addition of another polluter would also infringe in only a minor way. Once a river has been grossly polluted, for instance, additional pollution may not cause much more damage. It may be that there are thresholds that initially have to be overcome before pollution is perceived as a bad and once a certain further level has been achieved the environment becomes satiated with the pollutant and is indifferent to marginal additions.

In view of our conclusions on the ambiguity of society's perception of pollution and the firms' reactions to controls, there seems a strong case for ensuring that environmental protection is viewed in its local context as well as part of some grand national design; otherwise resources might be devoted to uses where society gets little real benefit. All the indications are that controls are becoming tougher and with the Common Market there is a pressure for uniform single standards that ignore the essentially situation-specific nature of the problem. More stringent control may be warranted in many cases but considerable research and investigation into pollution in its local context is necessary since single uniform standards ignore the economic and ecological realities of industrial life. Until recently pollution legislation has been framed in an essentially ad hoc way, reacting to short run pressures and choosing obvious and clearly definable targets that are not necessarily the most important. This has usually been on an industry basis; thus until very recently, the lead emissions from lead smelters were subject to much harsher conditions than the essentially similar lead emissions from secondary copper smelters. The primary aluminium industry has been subjected to enormous regulation whilst the state-owned steel industry continues to be a serious polluter. Different firms in the same industry sometimes come under entirely different control bodies; thus in our iron founders sample a few were controlled by the Alkali Inspectorate which in some cases was more lenient than the local authorities which controlled the rest of the industry, and local authorities' control was itself variable.

Pollution may be really like a balloon - squeeze it in one place and it extends in another. Thus, the firm required to install its own liquid effluent treatment plant ends up with an odour problem, another firm with an odour problem installs large fans to drive odours up a chimney and ends up with a noise problem, another firm burns its exhaust gases and ends up with a visual pollution problem from its chimneys, and so on. These difficulties arise partly because of the piecemeal approach to pollution control referred to and there is no overall pollution strategy and a multitude of bodies are responsible for control and enforcement. The differential control of the local authorities and Alkali Inspectorate are sometimes anomalous and the secretive nature of the latter precludes proper public debate and investigation of the costs and benefits of controls; a situation which leads to suspicion, misunderstanding and over-reaction by environmentalists. Thus we have experienced the opposite of pre-planned single standards but we believe that this also has been inefficient. The problem will only be properly tackled when there is complete co-ordination between the controlling bodies and the basis of control is the type of pollutant emitted and the damage caused rather than the industry-specific approach we now have.

Pollution Control Costs

In our research we were able to define essentially three types of pollution control cost. There were those which were instantly recognisable and represented the cost of fitting some device to a part of the production process but which did not directly impinge on that process. These end-of-line costs were those that we were able to get the most information on and are frequently the only costs that public policy makers seem to believe exist. Secondly there were in-plant costs which basically referred to the cost of an alternative production system foregone. Thus some processes or chemicals are not used because of their potential pollution load, even though they may be cheaper and more efficient than the ones actually being used. Sulphur dyeing is cheap but presents unpleasant problems to water courses; few firms in our sample used these dyes although many would choose to in an unregulated environment. Similarly, the current type of primary aluminium reduction process is used rather than a more efficient but more pollution-ridden one. Thus real in-plant costs may also include research opportunities foregone as well as current benefits foregone. In-plant costs are thus difficult to estimate and in our study we were only able to refer to them on a qualitative basis. Finally there are many intangible costs caused by pollution control. We referred in particular to space constraints faced by some firms which resulted from siting of pollution control equipment. Frequently pollution control equipment might take up valuable space which will preclude some production of goods, the breakdown of control equipment might necessitate a plant shutdown, or the installation of some in-plant control may result in lower total production efficiency, whilst substantial time and scheduling may have to be devoted to prevent the slightest risk of some materials being discharged to the environment. These all involve resources even though their costs may be difficult to trace.

For most firms in our sample pollution control costs represent a very small proportion of turnover and production costs. There are, however, certain instances where a combination of circumstances has resulted in pollution control costs considerably higher than the average and under certain financial conditions this can result in closures or mergers. The problems are more acute for older firms than for the newer firms; this is because the introduction of end-of-line pollution control techniques is always much more difficult in existing plant than in new plant. Further, new plant can be specifically designed to meet pollution objectives (e.g. noise) and thus, to an extent, pollution control costs may be "lost" in re-equipment. The large firm or the small firm which is integrated into a large group usually has an advantage because not only does it have access to greater capital resources but it is also able to utilise a pool of technical expertise not available to the small independent firm. Many of our respondents in textile dyeing and finishing emphasised this as important in their industry where the structural split is very clearly between small independent firms on the one hand and other firms making up large integrated organisations. It is especially important in the very small firms where there is less division of labour amongst management into their respective specialist skills; in these firms the manager deals with a multitude of functions and may be unable to cope with technically intricate areas in pollution control. Research associations frequently provide the necessary technical advisory service but we believe that they should be more actively supported by government aid to facilitate the task of research and information dissemination into pollution control that is, after all, the result of state intervention.

Small firms are also at a disadvantage because we found them frequently located in densely populated urban areas and not away from major urban areas as the larger firms tended to be.

It also appeared that pollution control was subject to considerable economies of scale due to the indivisibilities of plant. This may be a transitory problem since the pollution control industry grew serving the larger firms and its products became geared to these; in the longer term, however, it may adjust to the needs of the smaller firm. Our suspicions concerning the marginal costs of pollution abatement were confirmed by our own research and other published studies. The very sharply increasing marginal cost curve should deter the environmentalist lobby from pursuing their utopian objectives too far.

The costs of pollution control to industry are not, however, all negative. We have shown how unique situations like power and water shortages and periods of rapid inflation have led to a speeding up in the search for and communication of information on pollution control and recycling; at other times pollution control and charges themselves stimulate this search for knowledge. Substantial changes in output of pollutants can sometimes be prevented at near zero cost to the firm whilst the coincidental changes in the market and pollution legislation have made previously uneconomic waste control schemes such as energy conservation more worthwhile and many firms we visited are actively involved in waste management and recycling, although much recycling tends to be carried out by specialist firms because individual operators may lack the necessary capital, technical expertise and minimum quantities necessary to recycle their own waste.

There is not an infinite amount of slack available but certainly environmental controls have played a constructive role as a catalyst for some increases in firm efficiency in production as well as in pollution control.

So far pollution control costs in small firms have been low. The reason for this is the historical one that the degree of pollution control they have been required to undertake in the past has been slight because they have not been the obvious targets that large firms have been. Now, with pollution control regulations becoming more formalised, they extend down to the small firms and the latter are having to install control equipment at a time when profitability in industry is probably lower than it has ever been and when the cost of finance - when available - is high. These factors all add up to severe cash flow problems for small firms which may find themselves devoting a high proportion of total capital expenditure to pollution control rather than to expanding output; this is further exacerbated by the fact that the firm is being squeezed from all sides in all aspects of pollution abatement, viz. air, water, solid and noise pollution. If these measures had been required over a longer period of time they could probably have been accommodated.

It would be a great pity if the small and medium sized firms which provide flexibility and innovation should be threatened by environmental legislation.

The Incidence of Costs

It is important to take account of various structural factors of the industry in order to assess who ultimately bears the cost of control. In some industries costs of control are absorbed by slack resources but in others where there is severe competition or where the industry has been contracting, pollution control costs may present severe problems. The long run contraction of demand and the level of competition from both home and overseas limits the ability of the firms to pass on the costs of pollution control to the customer. Both the specific industries we studied fell into this category; the iron founding industry has been hit by substitution by non-ferrous metals and plastics and, whilst competition from imports is not of importance in this industry, it is naturally of great importance in the textiles industry of which dyeing and finishing forms a part. To an extent both these industries have also been deleteriously affected by a concentration of the buying power of their customers and this is particularly true of dyeing and finishing. The outcome of this has been an acceleration in the rate of concentration of these industries and, in the case of dyeing and finishing, a partial exporting of pollution problems to the new textile nations of the far east. Pollution control may also serve as a barrier to entry to new firms entering the industry. Thus, in the longer term, it might be a factor limiting domestic competition.

The Future

One of our main conclusions must be that some pollution is good for us, and yet some parts of the environmentalist lobby seem intent on pursuing polluters until the problem is completely abated (an impossible dream as well as an illogical one). We can only conclude that these people do not understand either the geometric relationship between pollution control and the costs of control, or that some valuable sections of the infrastructure of British industry can be done irreparable harm by their over-zealousness. This problem perhaps arises more than anything else from a failure of government agencies to follow the example of the US Environmental Protection Agency to analyse the functional relationship between costs of control and abatement of pollution on the one hand and the benefits and disbenefits that are entailed for society in the widest possible sense on the other.

Legislators should understand the short run problems of different sized firms at different periods of time. To some extent we have seen this in iron founding where there have been provisions to tide this important industry over the period when extensive pollution control equipment will be necessitated. Controls must be enforced but on a pragmatic basis, and the idea of single standards should be completely dispensed with since, if one conclusion of our research stands out above all others, it is that pollution problems, their costs to the firm and society are essentially specific to each individual case.

Changes in process and materials technology are constantly changing the nature of the problem. Initial observation suggests that these are primarily beneficial but essentially the problem of pollution is one of information and some changes which initially seem beneficial may merely contain hidden costs. Solvent dyeing may seem an answer to the large scale disposal of liquid waste from dyehouses but it replaces a high volume but largely non-toxic effluent with a low volume but potentially dangerous one. Only now are the fume dangers of electric induction furnaces being seen just as potentially hazardous as the dust and grit of the older cupolas. Few pollution solutions can be seen as final.

Ultimately, though, consumers and employees, as well as owners of capital pay for pollution; the "polluter must pay" principle should never omit this as a primary consideration whilst the pursuit of better quality final goods and services may entail higher pollution loads as well. Nothing is free, to anyone, and the environment is no exception.

APPENDIX 1

THE INDUSTRY QUESTIONNAIRE

The following pages show the questionnaire drawn up at the start of the project that was used as a basis for our interviews with businessmen. Naturally, for some firms particular sections of the questionnaire were more important than others; it is not possible, however, to show the full range of the supplementary questions that were asked in the interviews.

STRICTLY CONFIDENTIAL

INDUSTRY AND THE ENVIRONMENT IN THE EAST MIDLANDS

An Enquiry conducted by the Department of Industrial Economics, Nottingham University.

I 1. Approximately how many employees are there in your company?

Less than 50	50-200	200-500	500-1000	1000+

2. Briefly, what is the nature of your business?

....
....
....

3. Do you belong to an industry or trade association?

Yes	No

3.1 If yes, please name it :

....

——//——

II 1. Do any of your processes emit air pollutants?

Yes	No

1.1 If yes, under which of the following headings would you class them?

Carbon Monoxide	Particles and Dust	Sulphur Oxides	Nitrogen Oxides	Other (please specify)

2. What form of heating and power do you use?

	Oil-fired	Gas	Electricity	Coal
Heating				
Power				

2.1 Is your plant located in a smokeless zone?

Yes	No

2.2 If yes, has the introduction of a smokeless zone forced you to change any of your processes or plant?

Yes	No

2.3 If yes, would you give some idea of the cost :

....

3\. If no to Question 2.1, are you making plans to accommodate future air pollution restrictions?

No Need	Plans Beginning	No Plans

4\. Within the confines of your factory, do any of the production processes necessitate the use of dust extraction, face masks, or any other special equipment?

None	Face Masks	Dust Extractors	Other

4.1 What are those processes which necessitate this equipment?

....

....

4.2 What are the dust/fumes emitted from these processes?

....

....

4.3 Could you give us an approximate idea of the cost of this equipment?

....

4.4 Was this equipment introduced at the instigation of :

Trade Union	Management	Factory Inspectorate

5\. Does your company face any problems regarding air pollution that are not specifically covered by the Clean Air Acts?

....

....

....

....

—//—

6. Do any of your production processes discharge liquid waste?

Yes	No

IF NO, GO TO QUESTION 8

6.1 What sort of liquid waste is discharged?

Acids	Oils	Others (please specify)

6.2 Is any of the liquid waste biodegradable?

Yes	No

6.3 Is the waste discharged :

Direct to the Sewers	Direct into a river	Via a subcontractor

6.4 Is the waste treated in any way before it leaves the factory?

Yes	No

IF NO, GO TO QUESTION 7

6.5 In what way is the waste treated?

....
....
....

6.6 Was this at the instigation of :

The Local Authority	The River Board	Yourself

6.7 Could you give some indication of the cost of treatment of liquid waste?

....
....

GO TO QUESTION 8

7. Is the waste not treated because it is :

Not Harmful	The Costs are Excessive	Others (please specify)

8. Do you consider it possible for accidental discharge of effluent to occur?

Yes	No

8.1 Have you taken any steps to ensure that it does not occur - do you have any sort of automatic monitoring system?

No	Yes (please specify)

8.2 Has your company been prosecuted in the last five years for water pollution through effluent?

Yes	No

8.3 If yes, was this through an accident or because your monitoring system/staff failed to spot the trouble quickly enough?

....

8.4 Do you have any specific liquid effluent treatment problem which you feel someone at the University may be able to advise upon?

....

—//—

9. What solid waste results from your production process?

....

....

9.1 Are any of these recycled?

Yes	No

IF NO, GO TO QUESTION 9.4

9.2 Which wastes are recycled?

....

....

9.3 How is this done?

....

....

9.4 What happens to the waste not recycled? Is it :

Burned	Dumped	Removed by Local Authority	Removed by Private Contractor

9.5 If you do your own dumping, is this on :

Your own tip	Local Authority tip	Another

9.6 If your local authority helps you wish waste disposal, how do you rate the service?

Bad	Poor	Average	Good	Excellent

9.7 Is this service provided by the local authority free of charge?

Yes	No

9.8 In what respects, if any, do you think local authority services could be improved?

....

....

9.9 Could you give us some idea of the total costs of solid waste disposal?

....

....

10. Have your waste disposal methods changed since the introduction of the Deposit of Poisonous Wastes Act 1972?

Yes	No

10.1 If yes, in what way were your wastes disposed of before the Act?

....

....

....

—//—

11. Are there any areas within your factory where there are specific noise problems?

Yes	No

11.1 If yes, what steps have been taken to avoid these?

....

....

....

11.2 What effect have noise control measures had on efficiency?

Reduced	Increased	No Effect

11.3 Could you give us some idea of the cost of anti-noise measures you have taken?

....

....

——//——

III 1. Overall, what proportion of your production costs do you consider pollution and noise control measures account for?

Less than 5%	5-10%	10-20%	More than 20%

2. Do you think this proportion is likely to change in the future?

Yes	No

2.1 If yes, by how much?

....

3. Are you able to absorb these costs at all, or are they passed on to the consumer in the form of higher prices?

Absorb	Pass on

3.1 Could these costs possibly result in reduced employment?

Yes	No

3.2 If yes, could you give an indication of the size of this effect?

....

3.3 Do you think it is fair that industry should pay the whole cost of pollution control?

Yes	No

4. Do you think that the monitoring of pollution by local authorities and central government is :

Too Slack	Adequate	Too Tight

5. Do you think that existing standards are :

Too Slack	Adequate	Too Tight

6. Would you be prepared to see heavier penalties introduced for contravention of pollution regulations?

Yes	No

7. Do you think there should be a central government grant for the pollution control measures industry undertakes?

Yes	No

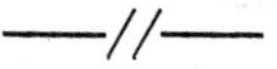